Technotrends

Technotrends

How to Use Technology to Go Beyond Your Competition

Daniel Burrus

with Roger Gittines

HarperBusiness

A Division of HarperCollinsPublishers

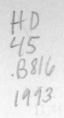
HarperCollins books may be purchased for educational, business, or sales promotional use. For information, please write: Special Markets Department, HarperCollins Publishers, Inc., 10 East 53rd Street, New York, NY 10022.

FIRST EDITION

Designed by Alma Hochhauser Orenstein
Illustrations by Tim Urban

Library of Congress Cataloging-in-Publication Data

Burrus, Daniel.
    Technotrends: how to use technology to go beyond your competition/ Daniel Burrus, with Roger Gittines.
            p. cm.
    Includes bibliographical references and index.
    ISBN 0-88730-627-6 (cloth)
    1. Technological innovations—United States—Management.
    I. Gittines, Roger. II. Title.
    HD45.B816 1993
    658.5'14–dc20   92-56234

93 94 95 96 97 ❖/HC 10 9 8 7 6 5 4 3 2

*To my father, an engineer, who fostered my love for science
and taught me by example the power of integrity and focus.*

*To my mother, who taught me the power of observation,
humor, and love.*

*To Patti, the love of my life, who has been at my side
as I gave birth to several businesses, lending support,
encouragement, and feedback.*

*And finally, to the thousands of people from my audiences
around the world who have taken action on the ideas I
sparked in their minds.*

Contents

Acknowledgments

Writing a book starts with a vision. To bring that vision into reality takes a team of very special and gifted people. I deeply appreciate the great effort expended on behalf of this project.

I want to thank Margret McBride, my literary agent, for having the vision to look beyond the manuscript and see the positive impact my message could have. In addition, I want to thank Winifred Golden, Margret's colleague, for her dedication and thoroughness.

Roger Gittines has been invaluable in helping me to focus the many concepts I share in my speeches and seminars into the confines of a single volume. Over the long months of research and writing, we formed a solid literary partnership that has been deeply satisfying to us both. Together we would like to thank our new friends at HarperBusiness and HarperCollins, especially Virginia Smith, our enthusiastic and supportive editor, Bill Shinker, Susan Moldow, Jack McKeown, Karen Mender, Mike Leonard, Connie Levinson, Lisa Berkowitz, Maureen O'Neal, and Bob Spizer, who shared the vision and focused their talents on this project.

I also want to thank Patti Thomsen for her countless hours of transcribing, editing, and feedback.

A special salute goes to Ken Blanchard, coauthor of *The One Minute Manager,* for his early advice and encouragement, and to all those who have been there when I needed them.

Introduction

A New Game

This book takes on a very large and important subject—more or less.

I'm not being coy. Its scope is nothing more, really, than our future: the way we will live, work, and make dreams come true. Nothing less than the rest of your life.

Those two points of reference, the one seemingly cosmic in its sweep and the other highly personal, are not at all far apart, although much of what is written about the future leaves that misimpression. There is a tendency among futurists to forget that while sweeping predictions and fantastic high-tech visions of tomorrow are good entertainment, they are typically far too general, making them of little use to those who want answers to a pair of age-old questions: What's *really* going to happen next, and what am I going to do about it *today?*

In fact, the practice of spinning impressively tangled technological webs and presenting often conflicting social trends serves to deepen the confusion and bewilderment that is a feature of the modern malaise that makes effective goal setting and decision making so difficult.

In the pages ahead, while you'll find much that is new and perhaps startling in terms of the latest technological breakthroughs, I guarantee that in addressing what's next we'll never stray from the effort to identify *what you can do about it today.*

And what can you do about it today? First, in a broad and general way, it is necessary to understand the nature of the profound technological changes that are under way. There's never been anything comparable in all of human history! Second, just as our ancestors—and we in turn—learned to use basic tools like hammers and crowbars, it is time to understand that no matter how complex the specific technology may appear to be, it too is a tool that must be mastered. We don't have any other option. And third, with new tools come new rules that determine how institutions and individuals function effectively.

The status quo has been shattered by technological change. As a result, success or failure hang in the balance for the remainder of the 1990s and the twenty-first century. Ignoring this reality would be tantamount to having shrugged off the development of water and steam power and the invention of the internal combustion engine. Many of those who did were left behind by the modern age, marooned in pockets of poverty and despair, deprived of the sustenance of the past and denied the promise of the future. Those who repeat the same mistake in our era are doomed to a similar fate.

Why a Card Game?

If you thumb through these pages to get a general sense of its themes and emphasis, you'll notice dialogue between characters who are playing cards. Odd, isn't it? A book about the future should be filled with interviews, charts and graphs and gadgets, trends and shocking statistics and predictions.

No, not really. The cards and the "game" that I have created around them serve as metaphors. I wanted to demystify the inherent complexity of modern technology. The card-playing analogy illustrates how *anyone,* regardless of technical background, can creatively apply technology to gain an upper hand.

America is a nation of gamblers. It's a trait we have in common with all countries and cultures. No matter what our nationality or place of origin happens to be, beating the odds is a universal and visceral challenge. For decades anthropologists have been finding primitive gaming paraphernalia in ruins of ancient settlements all over the world. Whatever the tongue, whatever the temper of the times, there are phrases that pulsate with life and sparkle with hope: "Let it ride," "Go for broke," "Bet on the come (-out)"—and they have equivalents in many idioms from Las Vegas to Angkor Wat. It's our human legacy. We take risks. We love to win and hate to lose.

The reception hall at Ellis Island was modern America's first great casino. In that echoing cavern of dreams many of our ancestors bet the farm, bet the factory loom, bet the ghetto walls; they bet it all and made history on the come.

Today, the children, grandchildren and great-grandchildren of a people who defied the odds are being told that the game is all but over: The wheel of fortune has turned against us.

Technotrends disputes that notion.

My premise is that the United States has always played by its own rules, essentially inventing an economic, social, and political "game," and teaching it to the rest of the world. All in all, we were very good teachers. Some of our more attentive pupils tore down the Berlin Wall, reunited Germany, scuttled Soviet communism, rebuilt Japan, and walked off with about 30 percent of the domestic U.S. car market.

Now, many of those same pupils are beating the United States at our own game. For many Americans, the objective isn't even winning anymore. It is a simple matter of survival, just hanging on. But this book is about winning, about re-inventing the future—a successful future. We can and we must go beyond an attitude that equates personal and economic success with managing to somehow "dodge the bullet" for yet another day. The time has come for America to once again play by its own rules—new rules—that will take us beyond the competition, and beyond the concept of competition itself.

We seem to be smothering in bad news and pessimism. I have solid reasons to believe, however, that the prophets of doom are wrong. The game is far from over. It has merely changed—changed in large part by our own success. There are new players, tools, rules, and strategies. We just haven't recognized them as such. It is time for the teacher to learn, a time to take risks, a time to win again.

Games Other People Play

For the past decade, as a professional technology forecaster and consultant to multinational and U.S. corporations, many of them Fortune
500 companies, national and international associations, universities,
and government agencies, I have been providing a new and powerful
method for both viewing and shaping the future that contains proven,
practical strategies and insights. Many of these organizations took
action, others did not. Up close, I've seen tremendous success for the
individuals and businesses that took action, and the damaging results
of failure for those that did not. Success, regardless of how you define
it, has little to do with how much you know about the latest technology. Rather, it is first linked to a steadfast commitment to applying
technology in new and powerful ways. Technology that sits on a shelf
gathering dust is useless. Would Wal-Mart be the success story it is
today if Sam Walton had decided to spend years studying EDI (electronic data interchange) instead of moving quickly to develop creative
applications and install the technology to link all his retailing outlets to
his suppliers? Certainly not! Sam Walton had only a rudimentary
knowledge of what was involved, but he knew enough to act. It is this
substantive, action-oriented experience that is the core of *Technotrends*.

Technotrends cuts to the heart of the issue. For decades, many
leading American companies were themselves beyond their competition but didn't know it—at least not in those terms. Recognizing that
they were up against vastly superior products and services, would-be
competitors wisely avoided head-to-head confrontations. Instead, to
avoid being crushed by the U.S. behemoths, our rivals gravitated
toward business opportunities that were not being pursued by those
firms. Out of necessity, they embraced new technology and new
methodology. This, in fact, is the classic first step to going beyond
your competition. As I have asked my clients many times, why compete when you can create a new market and have it all to yourself?

The advantages are obvious, and so are the consequences: A
whole new economic game evolved within the last decade. Roles have
been reversed. Those that had been virtually untouchable—giant corporations, national retailing chains, powerful trade unions, entire
industries that were once far and away beyond their competition—are

struggling desperately to survive as serious players. It has come as a
rude shock; frightening, disorienting, depressing. Still, I believe every
American is ready, willing, and able to master the new game now
being played with deadly seriousness by our rivals. It is a challenge
we face and must boldly confront as individuals, businesses, and most
of all, together as a nation.

What is needed is an easy-to-follow, straightforward overview of
how and why our world has changed and what we can do about it.
Ultimately, there is no better way to accomplish this goal than by
directly appealing to the gutsy, risk-taking qualities that are charac-
teristically ours. *Technotrends* does that by inviting the reader to
actually take part in the "game." Indeed, the metaphor itself gives the
book a unique format: a series of friendly card games, much like the
weekly poker or bridge nights enjoyed by many people. Each chapter revolves around an evening of play. I'll be there as host and
instructor to explain the basics, introduce the new cards that have
been added to the deck, and define the new strategies for winning—
ranging from when to hold 'em and when to fold 'em to trumping
the joker (i.e., government regulations, inefficiency, trade barriers,
etc.). Step-by-step, we will learn to go beyond our present and future
competition.

Characters You Can Relate To

I have created a cast of fictional characters to take part in the game, but
their backgrounds, personalities, and career profiles are drawn from
real life. There is a chief executive officer from a large multinational
manufacturing company, an inner-city high school teacher, a food distributor who owns a family business, a salesperson, a woman who has
just inherited the family farm, a manager who was laid off and recently
started a small business, and a hospital information manager. Each will
gain a major advantage by learning how to use technology to change
the rules of the game. My intention is to give the reader an alter ego
that will pull him or her into the story and into the future. I have drawn
the characters as inclusively as possible, because this is not a book
exclusively for corporate CEOs, small businesspeople, or white-collar
workers. The technological and organizational changes that are occur-

ring impact on all of us. Every individual, every business must learn and master the new game. There's no other option. It is the only way to regain lost ground and to stay on top in a world changing with such rapidity that one of the key new rules is *"If it works, it's obsolete."*

Shaping the Future

The pessimism that is paralyzing many societies, corporations, and individuals reflects our inability to answer the question I posed earlier: What's really going to happen next? As hard as we try, rapid change has prevented the future from coming into focus.

This book demonstrates that there are identifiable, cogent answers that are also provocative, stimulating, and immensely valuable. *Technotrends* is grounded in a new and proven method for shaping our future.

Over the past decade, I have delivered over 1,000 speeches throughout the United States, Europe, and Asia. The audiences range from small groups of about 100 to larger gatherings of 5,000 to 6,000. People continually ask me this question: "How can I turn rapid change into an advantage?" The answer, simply stated, is that the winds of change are not as important as how you set your sail. It is possible to navigate uncharted waters even through dense fog if you know where you want to go and you use the correct navigational tools.

Recent innovations in science and technology have provided us with a "new" set of tools to work with that will greatly increase our productivity and efficiency in all areas. Knowing what these "tools" are and how to apply them creatively is rapidly becoming a matter of business survival and a key to personal success. The new tools are actually creating permanent change so rapidly and so pervasively that the world of tomorrow cannot be the same as the world of today.

Of course, many people find the idea of greeting a new world every morning frightening and ominous. Some feel that change is out of control and that science and technology will ultimately destroy the human race. Others resign themselves to the belief that changes are coming upon us at such a pace and in such erratic spurts as to be totally unpredictable. They would argue that long-range planning is impossible because all an individual or corporation can do is react to changes as they occur.

But I don't agree! Even the most pervasive technological changes are both predictable and manageable. What's more, whether change is positive or negative will depend entirely upon how well we anticipate, understand, and prepare for it. The preparation process starts here in the pages of *Technotrends*.

Back to the Cutting Edge

Here's a checklist of what I set out to accomplish in this book:

- Equip you to recognize and master the changes already shaping the twenty-first century.

- Demystify and explain the technological revolution under way worldwide.

- Enable you to plan and develop a rewarding personal and professional agenda.

- Eliminate the uncertainty that breeds insecurity, frustration, passivity, and fear.

- Encourage people of all ages to understand and profit from technological change and to become excited and involved in building a better tomorrow by discovering creative applications for the new tools of technology.

- And finally, to introduce you to a uniquely personalized, tested system for shaping a future that can be everything you want it to be.

A Deck with Twenty-four New Cards

When it comes to the technology that will spawn the new products and services of the next ten years, the future has already been invented; the end products can't be bought off the shelf yet. The products and the services that will profoundly alter the way all of us live and work are more than pipe dreams—they are in the pipeline headed our way.

In the early 1980s, after extensively researching global innovations in technology, I was the first and the only science and technology forecaster to accurately identify the twenty core technologies that would drive the forces of change for the next twenty years. I created a classification system around those technologies that gave my clients an unprecedented ability to anticipate the historic transformation and position themselves for maximum advantage.

Technotrends is the next stage in the evolution of that classification system. Just as my clients have benefited from their awareness of the twenty core technologies, you will gain a decisive edge by being introduced to the twenty-four new tools for the twenty-first century. These tools have all emerged from the core technologies represented by my original taxonomy (classification system). These are the twenty-four new tool cards in the deck. With them the reader will learn to play a whole new game and play it to win.

Competers vs. Innovators

Historically, real successes have come when individuals and companies have focused on leadership in the marketplace and teamwork internally.

This book is designed to enable you to move beyond your competition by nurturing, promoting, and enhancing innovation and original thinking—both individually and within your organization. My goals are to stimulate you to become an innovator and to show you how to constantly go beyond those whom I call the *competers*.

What's the difference between *competers* and innovators? No, *competers* is not a misprint. I'm taking a liberty by creating a term for those who reflexively compete rather than seek to gain a strategic advantage through innovation. Here are some of the distinctions you'll discover in the following pages:

- Competers tend to copy what others are doing; innovators are constantly looking for better ways of thinking and acting.

- Competers get locked into set patterns; innovators constantly cultivate a creative mind-set.

- Competers seem to believe the future will take care of itself if they take care of the present; innovators are focused on their future goals and building a path to get there.

- Competers tend to see scientific and technological developments as threats to their status quo; innovators focus on how they can apply new technologies to open up new opportunities.

- Competers tend to collect and swim around massive amounts of raw data; innovators look for ways to translate raw data into useful information.

- Competers tend to only react to trends; innovators learn how to predict and even create trends and profit from them.

- Competers take a short view of planning and consider it a necessary evil; innovators take a strategic view of planning and know the value of building change into the plan.

- Competers dread change and resist it as long as they can; innovators seek to remain adaptive and to master change.

- Competers often avoid anything that would cast them as being significantly different from their competitors; innovators seek to maximize their differential advantage.

- Competers in management positions try to control and direct their people; innovators seek to empower their people for positive action.

- Competers often complain about how unproductive their people are; innovators realize that people are their most upgradable resource and constantly look for ways to help their people to be more productive.

- Competers often think about how they can use high technology to cut their work forces and save money; innovators seek to integrate strategy, technology, and people.

- Competers are often big on standardized operations that force people to act in predictable ways; innovators encourage creativity in their people.

- Competers are annoyed by problems and see them as enemies of progress; innovators go looking for problems they can turn into opportunities.

In short, competers are usually so caught up in meeting their day-to-day challenges that they can only worry about the future, while innovators see the present only as a stepping-stone they can use to get to a bigger and better future.

Shaping Your Future

Can innovators predict the future? No, not in the sense of gazing into a crystal ball or using tarot cards. Many surprises are still in store for us. But anticipating broad patterns of change as well as foreseeing specific shifts in the ways and means of doing business and living our daily lives, is possible. It is important to recognize that the most sweeping and pervasive permanent changes in history that were predictable— I'm thinking of events other than acts of God like shifts in climatic conditions or random chance—have grown out of scientific and technological advances. The fact is, science and technology don't merely reflect permanent change, they *create* it.

During times of relatively slow change, there were many acceptable methods for predicting the future with some accuracy. A rapidly changing world has, however, rendered most of them far less effective and in some cases obsolete.

Often, well-meaning analysts misread the future because they rely on what I refer to as soft trends. And I should note that no other forecaster or futurist makes a distinction between soft trends and hard trends. However, it is of critical importance. Soft trends contain possibilities usually in the form of best-case and worst-case scenarios— speculation on what might be. They are usually based on public opinion surveys, news events that show a pattern, shifts in popular culture, and anecdotal flotsam and jetsam. Hard trends, as the name implies, are concrete and verifiable. They are based on the scientific study of tangible, physical, existing items in our world. They are not derived from shifting patterns of taste and fashion. I'll give you an example: Studies of the aging population of the United States have revealed very accurate hard trends. People are tangible and their age is measurable, revealing far more accurate forecasts over time. By choosing *Technotrends* as the title for this book, I am declaring that the ultimate hard trends spring from technological change.

In the chapters ahead, you'll be introduced to the hard trends that shape the future with all of their ramifications, challenges, and thrilling possibilities. I believe that the golden door has swung wide open, creating a time of unprecedented opportunity for any individual who notices that the way is clear and walks through with the right knowledge. I also believe that this is the greatest era ever for people who want to run successful businesses and other enterprises.

I like to think of the twenty-four cards that science and technology have dealt to us as incredible new tools we can use to shape our futures into anything we want them to be. I hope you are as excited as I am about learning how to go from surviving to thriving in the future. After all, that is where we will all spend the rest of our lives.

Join me as *Technotrends* takes you on an intensive search for personal and corporate winning strategies that will put you beyond your competition.

The Players

E. Maurice Andrews Chief executive officer (CEO) of a major mainframe computer manufacturing corporation.

Samantha Brewer A twenty-four-year-old independent life insurance agent.

Lonnie Weineck A former middle manager from a Fortune 500 company who has recently started his own landscape design firm.

Brenda Jacobs An information manager in a large city hospital.

Glenn Clayton Proprietor of a family-owned food products distribution company.

Doug Stedman An inner-city high school science teacher.

Tanya Conte A dairy farmer who inherited the operation from her father.

1

A Reality Check

Beating the Traditional Players

E. Maurice Andrews was the first to arrive. I was arranging the chairs around the table for the evening's card game when I looked out the dining room window and saw a long black limousine pulling up in front of the house.

Maury didn't get out right away, which gave me a chance to take one last quick review of the business executive's background. As the chief executive officer of a giant multinational computer manufacturer, he was accustomed to seeing his name on the pages of *The Wall Street Journal, Who's Who,* and the invitation list for White House state dinners. I wouldn't call it slumming, but he certainly was traveling outside of his usual circles by attending a series of card games with a high school teacher, a farmer, a hospital information systems manager, the founder of a struggling landscape design firm, an independent insurance agent, and the owner of a small food products distributorship.

Maury was there because he was a desperate man. Social cachet, or the lack of it, were of no importance. His company and his career were in deep trouble. Maury had heard about my unusual card games from a fellow CEO who had attended the month before. The two men belonged to the same country club, and while they worked in different industries, they shared many of the same problems. They often commiserated and exchanged business horror stories on the golf course.

"I don't have time for cards," he gruffly told his friend (who gave me a full report the next day). "I barely have time for golf. I've got a corporation that's sinking like a stone and you're talking about cards."

"Forget about poker and bridge, Maury," the other CEO advised as they arrived back at the clubhouse after eighteen holes, one of the few chances Maury had had to unwind for several weeks. "Dan uses cards as a metaphor, a teaching tool. Think of it this way. Imagine yourself sitting down to a high-stakes poker game. The other players play every night and are all very good. You, on the other hand, have never played the game. You're in deep trouble! That is, unless you get to provide the deck of cards. And, you get to define the rules of the game. You now have the ability to turn their advantage into a disadvantage and put yourself beyond the competition. Using his card metaphor, Dan will teach you how to use the new tools of technology to change the rules of the game. You'll learn more about how to turn rapid change into a competitive advantage in a few nights at Dan's place than in two months of high-level seminars."

I had been warned that Maury Andrews could be cantankerous. But his admirers said most of it was a product of stress and anxiety generated by his company's recent poor performance. For decades one of the industry leaders, a textbook example of excellence, it was hemorrhaging badly. The bottom line was showing a loss for the first time in history. The work force had been reduced by 15,000 people, and another round of downsizing was likely if business didn't start reviving soon.

Yes, Maury was a little tense when he knocked on my front door.

"Come on in," I said. "You get the prize for being first." I introduced myself and showed him into the dining room. One of the things I liked about Maury was that he was very direct. He immediately told me that he was skeptical about whether he would derive any benefit from "fooling around with a bunch of cards," as he put it.

"Okay, Maury, to heck with the cards. Before the others get here let's talk about the effects of fuzzy logic technology on your company."

The CEO gave me a blank look. He glanced around the room. "Nice house, Dan," Maury said, as if he hadn't heard my comment about fuzzy logic.

I offered him a chair. "We've got a couple of minutes. Are you offering your customers neural network capabilities?"

Maury sat back and crossed his legs. "We make and sell mainframe computers; that's our bread and butter. Fuzzy logic and neural networks are specialized technologies that a few of our customers might be using. My engineers and R&D people keep up with topics like this, while my focus is leading the company. I didn't go to engineering school, I went to business school."

"I know, Maury," I said. "Harvard MBA, class of 1964. My point is that simply by mentioning things like fuzzy logic and neural networks, I provoked in you a classic case of executive technophobia—'I didn't go to engineering school, I have my engineers and R&D people deal with topics like that.'"

"Well, I didn't go to engineering school and I do have my technical staff evaluate emerging technology."

"You don't have to be an engineer to play the game," I said. "The cards you will be using tonight will give you a short course in the kinds of technology that are in the process—and I mean right this minute—of changing the way you and your competition will do business.

"Let's face it, your technical people were aware of parallel processing computers in the mid-eighties, weren't they?" Maury nodded yes. "Yet you, the leader of the company, were so focused on the success of your mainframe systems that you didn't pursue this new technology, while other smaller start-up companies did and in the last few years have been using this technology to steal your biggest customers."

"You've made a good point, but cards . . . ?" he started to interject.

"The cards demystify the technology. Almost everybody knows how to play cards. It's no big deal. There are winners and losers, strategies, techniques, skills, planning."

"Yeah, but fuzzy logic and neural networks are pretty damned complex. When my researchers begin to explain the principles, it gets boring fast," Maury said, checking the dial of his wristwatch.

"You would get bored fast if they tried to explain the scientific principles of that baby over there." I pointed to the telephone on the table in the hallway. "You use that device dozens of times a day and have only the vaguest notion of the exact physics of how it works. Why? Because you don't need to know the physics of a telephone to use it. You only have to know that it exists and then creatively apply it to what you are trying to do. By turning the new technologies into a deck of cards, I'm making them user-friendly—as familiar and non-threatening as . . . a telephone."

At that moment Doug Stedman arrived. He is a high school science teacher. In short order all the others were also sitting around my table, and the game was ready to begin.

The Future Isn't What It Used to Be

I put two decks of cards on the table. "I'm going to deal to Maury from this one," I said, pointing to the deck that had a green filigree design on the back of the cards. "The rest of you will get cards from this other red deck." I snapped the edges of the cards as I shuffled the deck.

I gave Maury one card facedown and then dealt the others a card each from the red deck, also facedown. I repeated the process, but this time the cards were faceup. Maury had a queen showing. Everyone else had cards of minor denomination and no face cards showing.

"Time to bet," I said. "One-dollar minimum, and whatever Maury bets—he goes first under the rules of stud poker, since he has the most valuable card showing—the bets double with each round." The group nodded in unison. "Also, in my games nobody drops out or checks." I saw Lonnie Weineck gulp. *Check* is a poker term that essentially means staying in the game without putting a bet in the pot. My rules meant that the players were totally dependent on the cards without any possibility of limiting their loses before the end of the hand.

Maury put two chips worth one dollar each in the middle of the table. The other players followed suit. I dealt out another round of cards faceup using the two decks. This time, Maury drew a king. Again, the others received minor cards.

After glancing at his hole card, Maury's eyes twinkled and he tossed out four chips. "I don't think I'm going to enjoy this game," Tanya Conte said as she matched Maury's bet. She had inherited her family's dairy farm three years before, and cash had always been in short supply.

When I had dealt out all five cards, with a betting interval after each round, it was obvious that Maury was about to clobber the rest of the table. He had an ace, a queen, a king, and a ten showing. They were all of the same suit.

The others didn't even come close to winning combinations. The pot was worth about two hundred dollars. Maury turned over his hole card, the one that was facedown, and it was a jack, and another spade like the rest of his cards. He had a royal flush, the highest hand possible.

Maury chuckled as he collected his chips. "And to think I almost didn't come tonight," he said with glee.

"I hope this is going to be an early evening," Tanya said. "I've got to be up early to do the milking."

"Me too. Got an early appointment," Samantha Brewer chimed in. She was the youngest player and seemed slightly intimidated.

"Don't be in such a hurry. Your luck might change," I said. But their luck didn't change. In the next hand Maury won again with a straight flush.

Glenn Clayton, the food products distributor, announced that he was "done."

"Amen," said Brenda Jacobs. "I can have my brains beat out at the hospital any day of the week." Lonnie and Doug were silent.

To avoid a mutiny, I suggested that we take a break from cards. I poured cups of coffee and eased the tension with small talk. Lonnie loved to discuss landscape design, and I got him to explain why the leaves of my lilac bushes were covered with a fine white powder. He said it was because of the unusually hot and humid summer. Before long everybody was talking shop, except for Doug, who made it a point never to bring his "office"—an inner-city high school—home with him.

After about ten minutes, I summoned the group back to the table. We had a brief discussion of the parallels between the game of poker, or any card game, and the game of life. All agreed that there were many similarities, especially when it came to knowing the rules and knowing how to keep score.

"Okay, before we start again, I think it would be a good idea to examine these two decks," I suggested. "I've been dealing Maury's cards from this one." I turned the deck faceup and spread the cards on the table.

"Holy cow," Tanya exclaimed.

"Well, you're the expert on livestock," I said. Everybody laughed.

The deck was composed of all the high-value, winning cards. Furthermore, it was stacked in unbeatable combinations. I turned over the other deck to show them that it contained nothing higher than a seven.

I told the losers that I had deliberately sandbagged them to demonstrate that under the rules of the *old* game Maury was a guaranteed winner. "Look at it this way," I said, "his company has a name that is recognized worldwide. Its reputation is first-class. The banks and the stock market are always ready, willing, and able to meet Maury's capital requirements. On top of that he has had more ready cash than he knows what to do with." I picked up a handful of his chips and let them cascade down onto the table.

"There's a massive customer base built up over the years, and those customers have very high expectations. Maury's operation comes equipped with a mammoth infrastructure, lots of networks, and that, in and of itself, generates a perception of permanence." Maury nodded his head in agreement. "He won," I said, "because he has always been a winner." Another nod from Maury, and then I surprised him—"Until now." He flushed slightly at that.

I dealt out another hand of cards, using the same decks. Once more, Maury had all the winning cards. As he reached for the pot, I stopped him.

"Not so fast. You lose, Maury. That four of a kind is topped by Samantha's straight 2, 3, 4, 5, and 6."

Maury knew his poker rules and immediately protested. He said that a four of a kind beats a straight every time. "Not any longer, Maury. The rules have changed.

"There's never been anything like it. The emergence of new technology is driving change so fast that change itself has changed. We've lost most of the traditional benchmarks that have been used to evaluate our circumstances. As recently as ten years ago, a traditional player like GM, GE, or IBM could look down the road and expect tomorrow to look pretty much like today. Not anymore.

"The uncertainty of this maelstrom of change was best captured by a famous statement that has been variously attributed to the French poet Paul Valéry or—from the sublime to the ridiculous—baseball great Yogi Berra: 'The future isn't what it used to be.' No matter whether Valéry or Berra was the author, the observation is both a *bon mot* and a home run. Indeed, the future is not what it used to be."

If It Isn't Broken

"From here on, Maury," I said as I glanced around the table to make sure every person in the group knew that I wasn't just talking about corporate CEOs, "you're playing a game without knowing the rules."

"Hello, sucker!" Glenn Clayton interjected.

I nodded. "Exactly. Who'd be crazy enough to play a game without knowing the rules? Or, just as bad, playing with a rule book that has gone out of date? It would be like an American League baseball coach stopping in the middle of the seventh inning and asking, 'Where'd that guy come from?'" I did my best imitation of Yogi Berra.

"That's their designated hitter, coach."

"'The designated what . . . ?'"

"I'll bet there are still some baseball fans who haven't caught on to the designated hitter," Brenda Jacobs observed.

"That's probably true, and why?" I asked. "Some die-hard fans expect baseball today to look like baseball yesterday and baseball tomorrow. And it—baseball—usually does." I continued: "Now, let's talk about real life. If you expect tomorrow to look like today, it is only natural to keep doing what you're doing," I said. "Particularly, if you have been immensely successful. In fact, Maury, I'd be willing to bet that you have said these words more than once: 'If it isn't broken why fix it?'"

"Yes, I have, and it makes a lot of sense," Maury said.

"And you've probably been saying it," I persisted, "since you became CEO in the mid-1970s, right up through the late 1980s at about the same time your larger customers discovered the power of parallel processing computers, and when your smaller customers started realizing that cheap and flexible networked PCs could do many of the same tasks as the mainframes you manufacture. Right then and there, 'it' was broken and you didn't even know it. The company was so big and rich—remember those were winning cards in the old game—the loss of small increments of your market share was no big deal. There was plenty more where that came from; once the customers realized the limitations of the PC, you figured that they'd return."

Maury looked uncomfortable. "What you are saying is," Maury volunteered, "we got too fat, dumb, and happy." Before I could answer Doug Stedman spoke up.

"Our school system used to share time on the city's mainframe. PCs are everywhere these days, and we virtually quit using the mainframe a few years back. Come to think of it, I'll bet your outfit's logo was on the hardware."

"It's too sophisticated and expensive for school systems anyway," Maury said defensively.

Doug pressed on. "What Dan said about the customers returning once they discovered the PC's limitations just didn't happen. Those early PCs were slow and didn't do much except sit there and blink. Every six months or so, though, another hardware generation and software upgrade came out. Now it's the mainframe that's showing its limitations."

"Not if you are launching satellites," Maury replied.

"I'm trying to teach seventeen-year-olds, and there are a lot more of them than satellite launches."

Things were getting a little testy, so I jumped back in. "Traditional players, playing with the traditional rules, assume that change will occur the way it always has—slowly. Maury, you probably figured in the mid-1970s that mainframe technology would hold its edge for at least another thirty years. More than enough time for your R&D people to come up with improvements to keep you ahead of the competition. Besides, in thirty years you and the rest of the senior management team would be on the verge of retirement." I paused to let him react.

"I can't argue with that, Dan." He took off his wire-rimmed eyeglasses and examined the lenses for specks of dust. "After all, next to IBM we practically invented the mainframe. It took about thirty years to get the bugs out. Few were even close to us in terms of manufacturing capability, patents, the distribution and service network. With an edge like that, thirty years was a safe bet. But it was not as if we were just resting on our laurels. I built a world-class corporation in the meantime." Maury put his glasses back on and looked around the table as though he expected someone to challenge the statement.

"Indeed you did," I said. "The only problem is that you built that world-class corporation around a product that would be in decline much earlier than you expected, and today you are paying the price."

The Past Tense of Blockbuster Is . . .

I got up and walked across the room to the television set. "Let's consider another industry altogether—the video rental business. Blockbuster has it made. They have locations scattered all over the major cities and suburbs, a superb inventory, and a solid customer base. But if Blockbuster continues to use its current success strategies, it will be finished in five years. Blockbuster to Blockbusted. Why?"

Brenda Jacobs spoke up. "At the hospital we have some patients who try to call out for pizza deliveries. Of course, we stop them at the front desk. But I've noticed that the same thing is happening informally with videotapes. Visitors routinely bring them for patients." Brenda shrugged. "If Blockbuster isn't careful, it will be done in by the same thing that is destroying little neighborhood pizza joints. Home delivery."

I gave Brenda a thumbs-up and said, "Right. Domino's and Pizza Hut bring the product right to your front door. No more getting in the car and going to pick up an already cold pizza. Well, the technology exists right now to bring the contents of the entire store to our homes. No more getting in the car and driving to the Blockbuster outlet. And I don't mean that the answer is to hire teenagers to zoom eighty miles per hour down one-way streets to beat the clock."

"I think my kids spend more time going to and from the video store than they do watching the tapes," Glenn Clayton said. The others nodded in agreement.

"You're right," I agreed. "Surveys show that the average customer spends more time in the video store, and going back and forth, than he or she does watching the tape."

I picked up the remote control device from the top of the television console. "Soon, your kids will use this instead. They'll tune the set into what will look exactly like the interior of a typical video store. They'll browse around until something looks interesting. *Click, click, click*. There's a shelf with all the Terminator films. Let's take a closer look. *Click*. How about *Terminator VI*, I haven't seen that yet. *Click, click*. Up comes a preview. Looks pretty good. *Click*. It's rented, billed out to their membership number and a display comes up asking if they'd like to see it right now or wait until later. When *Terminator VII* is released, they will be so advised because the system remembers that they rented *Terminator VI*, and it will reserve the video for Thursday nights because that's when they usually watch a movie."

"Forget Big Brother," Lonnie said. "Big Blockbuster is watching!"

"George Orwell wrote *1984* almost fifty years ago, and he was a pretty good science forecaster," I said. "He could see the technology beginning to emerge that would allow for sophisticated two-way telecommunications. It started off as pure science fiction. And it is reality today. If Blockbuster decides that it's got a lock on the market, they will find that the lock is about to be picked by somebody else using this new technology. The retail locations, the inventory, the customer base won't matter in the long run. In the past, the customer went to the shop. In the near future, the shop will go to the customer."

I moved back into the dining room and put my hand on Maury's shoulder. "One day soon, just like Maury, the CEO of Blockbuster is going to sit down at the card table, pick up a hand of traditional winning cards, and start losing—and wonder what's happening.

"The rules have changed—that's what's happening."

Permanent Change—This Too Shall (Not) Pass

Glenn, Brenda, Samantha, Lonnie, Tanya, and Maury all made the same demand simultaneously: "Give us the new rules!!!"

"Okay, here's one: *Past success is your worst enemy,*" I said.

Glenn Clayton had an immediate rejoinder: "So we're all supposed to strive for failure?"

"No, but you've reminded me of another rule: *Learn to fail fast.*" I paused while some started jotting down the new rules. "I'll come back to the one about learning to fail fast. First, I want to explain 'past success is your worst enemy.'

"Maury has made a huge investment in his company's success. Naturally, he doesn't want to walk away from it. But that instinct puts him at a disadvantage during times of rapid change—and change has never been as rapid, as relentless, and as all-pervasive as it is today. What happens is that we hunker down and circle the wagons to protect our profit centers—our cash cows—from the effects of change. The advertising budget is tripled, the marketing staff gets strict marching orders, the sales reps are slapped with higher quotas.

"In the short run there might even be an uptick in sales, which seems to validate what management was thinking all along: 'This too shall pass.' They are making a typical mistake by assuming there is only one kind of change that they need to worry about. And that is cyclical change.

"The textbooks tell us all about cycles. First, you go up. Then, you come down. If you're lucky, you survive that ride and go back up again. Managing the cycles becomes the name of the game.

"What is happening to Maury doesn't have anything to do with cycles. It is a second type of change—permanent change. Just as the Blockbuster customer, after getting a taste of home video delivery at half the price, will never return to the retail location, Maury's potential customers—the start-up companies that went with the cheaper, more flexible networked PCs—have developed a corporate operating life-style around the PC that is totally alien to mainframe technology. As they mature and expand, the tendency will be to turn to those suppliers who have been meeting their demand for 'new solutions.' Thus, the old players are virtually stigmatized for their past success. The question is—what have you done for me lately? The answer is—nothing.

"Furthermore, the large customer base built up over years of success is also a liability. These customers are wedded to an old generation of mainframes or other less-flexible, expensive-to-maintain, and, in many cases, outmoded technologies. They have to have their hardware systems updated, as well as supplied and serviced, which keeps the manufacturer focused on old technology that consumes available resources and often slows new product advances."

Lonnie broke in to say, "It sounds like what happened to the milling machine company where I worked for fifteen years. Their customers were all old-line operations that only wanted another copy of the previous model that they purchased ten years before. We took a beating on anything new or different, and after a while management refused to take any risks."

"Let me guess what happened next," I offered. "The Japanese came along to offer a similar but more flexible machine at a lower price to the old-line customers, and new and different equipment to the smaller one- or two-man shops that were beginning to nibble around the edges of the business."

"You got it," Lonnie said.

"They could afford to do that because success was not yet their worst enemy. They didn't have a huge financial and emotional investment in the old milling machine. They could clone it, attach a few bells and whistles, lower the price, and treat it like an interim technology while they introduced the new stuff to little companies that were growing by leaps and bounds. They didn't have a cash cow to protect."

"Without a cash cow, Dan, there's no cash," Maury said. "We've been introducing new products, but the mainframes still supply the bulk of our revenue and profits. I spent almost four years trying to get a version of a PC going. It was like pounding my head against a brick wall. The margins on PC hardware are so thin, it has been a money loser," he complained.

"Back to that other rule—learn to fail fast," I replied. "It used to be that a wealthy individual or a well-capitalized company could ride out a down cycle with heavy, money-losing investments, and then make it back on the other side when the cycles changed. These aren't cycles anymore.

"Look at it this way, Maury. The mind-set you developed as a successful businessman will lead you to hang in there with your focus on mainframes as your main source of income. The CEO of a start-up operation without your history, resources, and large customer base would take the hit, see the potential pitfalls, and dump out. He would fail—fast. You, on the other hand, fail but fail slowly. In the long run, the hemorrhage is more costly and prevents you from focusing on more promising opportunities."

Tanya Conte, the dairy farmer, had a question: "How do you know when to give up and try something else? Maybe you're quitting just as it's about to take off?"

"A new product—and it may be a good one—that needs six months, a year, three years to get off the ground is obsolete the moment it gets airborne. The company that just sold you the new product already has the next generation working. Today's competitors are getting new technology off the drawing boards and into application faster and faster. You need to assume, in any case, that if a product or a technique works, something more advanced has already been invented.

"Now, before we play another hand, I'm going to give you a summary of the new rules to demonstrate to Maury that if the old players learn the new rules, they can start winning again and winning big.

"Here are the new rules so far, and I will start with the one I just alluded to:

If It Works, It's Obsolete

Past Success Is Your Worst Enemy

Learn to Fail Fast

"I'm sure these rules make everybody here feel insecure," I said after jotting down the key points on a flip chart.

"You can say that again," Glenn Clayton sighed.

I nodded in encouragement. "There doesn't seem to be any fixed reference point. Particularly so when success itself can lead to failure. Well, there is one bright star to steer by, and I will display it in its own private box:

Make Rapid Change
Your Best Friend

"These rules are not optional. They are in effect whether we like it or not. The key is to recognize them and take action. Pretending otherwise doesn't do any good.

"In most cases, it would be a lot easier if we could use the old rules and assume that over the course of one's career change would be the exception. Once, and not long ago, several generations would go by without major alterations in the way our ancestors lived and worked. Even the major innovations of the Industrial Revolution took years to spread. Farmers, craftsmen, and other workers could continue to rely on methods that were passed along from father to son. Once an apprenticeship was completed, most of the skills that had been learned could be counted on to last for a lifetime.

"Unfortunately, it doesn't work that way anymore. I think of it in terms of a great river that is literally carrying change around the world.

No place is immune. The traditional definition of security—'I'll learn a skill, get a job, and stay put for the rest of my life'—no longer applies. The exciting thing is, though, fundamental change brings with it unparalleled opportunities.

"It's an optimist's dream come true. Even ideas that failed in the past can be tried again using the new technology to supply the missing ingredients until successful. The pessimists and the champions of the status quo are the ones who have the most misgivings about change. I'm alarmed when people tell me that they are nervous about the future. To me the sixties, the seventies, and the eighties were not a golden age by any means. Remember the rigid hierarchies, energy shortages, high interest rates, strict union rules? Nothing could be more stultifying than more of the same. I'm ready for the twenty-first century."

"I would be too if I were Samantha's age," Brenda said. "She's only twenty-four."

I shook my head. "Age has nothing to do with it. Think of it as the *ultimate* democratizing force. Today's traditional players—the big winners—may be tomorrow's losers." Maury frowned at me as I continued. "Every day is, in fact, a new day. Those of us who made mistakes yesterday get a second chance under a whole new set of conditions. Those of us who were financially disadvantaged now have a level playing field. Age is a barrier only if you erect it yourself.

"In some ways, Thomas Jefferson was 200 or so years ahead of his time when he wrote 'All men are created equal.' In the past—in Jefferson's day—those who captured control of territory, resources, or huge pools of capital were in a position to call the shots. They were *more* equal than everyone else. This old rule has been true for what seems like forever. Not anymore. Technology is evolving with such rapidity that any miser who attempts to hoard the latest developments and corner the market runs the risk of possessing a rundown antique shop, albeit one with a listing on the New York Stock Exchange. Maury and I both know that mainframe sales and service will continue to yield profits, but in a very mature market, using a mature solution. With organizational decentralization as an ongoing business strategy for large organizations, large mainframe systems will play a different and shrinking role. Will the margins be thinner and the competition rough for the remaining market? You bet!"

"I've got it, Dan," Maury said. "I'll rename the company Ye Olde Mainframe and Quilt Company."

We had a good laugh, and I dealt out another hand of cards. The players looked a little more confident. They didn't know all the new rules yet, but it was a start. Most importantly, they all realized that there was a lot to learn and a lot of opportunity.

2

Turnaround Time

Beating the New Players at Their Own Game

Maury Andrews was down to a few last poker chips, and he was worried. The brash self-confidence that he had shown the night before was quickly ebbing away. He was back to looking like the harried chief executive officer that he, in fact, was.

"Three kings," he said, exhaling in relief.

I pointed to Doug. "A pair of fives; Doug takes the pot."

"How's that?" Maury asked sourly as he tossed his cards down.

"As I've been explaining for the last couple of hours, the game has changed because of the new rules. Three of a kind used to beat a pair. Today, those three kings of yours mean that the old rule—bigger is better—doesn't hold anymore."

"I get it now," Maury said. "We're playing lowball poker. High cards lose, low cards win." He folded his arms across his chest. "I can't win because I'm being dragged under by my own history of success."

"Not exactly," I answered. "In what's known as lowball poker the pot does go to the worst hand. In this game, being dealt cards that won or lost in the past is meaningless." We had switched to a version of draw poker that allows the players to discard any of their cards and receive replacements from the dealer, but Maury was still trying to win with the face cards. "You had a pair of eights in your hand, Maury, and those would have beaten Doug's fives. Yet, you played the kings."

"It's hard to tell which new rules apply to which cards," he complained. "All we know is that there are new rules."

"Now we are getting someplace," I announced. "Even though he is aware that there are new rules, Maury still instinctively goes with what he is most comfortable with—the old rules."

"He uses what works," Samantha said, giving Maury a helping hand.

"Worked," I corrected her. "Past tense. He's assuming that the present is a dressed-up version of the past, but it isn't. The old rules *W-O-R-K-E-D*"—I spelled out the word emphasizing the *E* and the *D*—"because they were compatible with the old tools.

"Go back to the example of Blockbuster. If the present is like the past, the smart thing to do is go deeper into debt to buy more real estate for a hundred new retail outlets or buy related businesses. In the past we didn't have access to delivery systems such as fiber optics, direct broadcast satellites, and digital microwave transmissions to the home, and there was no receiving system like digital interactive television. However, now we do. Together, they allow us to create a 'virtual' retail outlet without an investment in bricks and mortar and mortgages. Video rental stores like Blockbuster could easily become virtual retail outlets that could exist simultaneously in millions of homes at the flick of a wireless television channel selector. Since their product, movies, can be stored in a central computer system and sent to the customer in digital form, there would be no need to buy or stock videotapes. Think of the savings in overhead expenses!"

I pointed to the flip chart in the corner of the dining room that still displayed the four new rules that I had presented soon after the game started. "We have new rules because there are new tools. In this case, digital interactive television is the new tool."

As had happened when I first introduced the concept of new rules, the group started clamoring for a list of the new tools.

"Not so fast," I said. "I've found that there is a tendency to go overboard about the latest gadgets and lose sight of the fundamental changes that are taking place all around us. First, it is necessary to understand that *technology alters reality*—alters the way we think and act. The invention of the campfire, and that's technology at its most basic level, turned humans into social animals. On cold nights, the warm cheery blaze drew our ancestors out of the dark forest into a circle of companionship. It may have taken decades or centuries of near silence before someone started to sing or dance or tell stories. But the reality of solitary hunters and foragers was nonetheless altered forever, and we are still gathering around the flickering modern campfire—the television—to listen to stories and to gain a sense of global community.

"In the end, it is not as important to know what to do with technology as it is to know what technology does with us." I paused a moment before continuing. "Therefore, the place to start is with the new rules, which are the footprints—the pathway—left by this process that alters the way we think and act."

Doug Stedman spoke up next. "I think I know what you're getting at, Dan, but aren't you reversing the normal cause-and-effect relationship? Shouldn't we be dealing with the cause first?"

"I am reversing cause and effect. It is the effect that has the deepest implications for success and failure in the future. The technological cause is important—and we will get to that—but technology is evolving with such startling rapidity that if you get fixated by today's big breakthrough, there's a good chance you will miss what happens tomorrow and the next day. If I can get you to accept that fundamental, rapid change is inevitable and welcome, that attitude will equip you to prosper in the twenty-first century no matter what new gadgets are introduced to our homes, offices, and factories."

Another Deck

I gathered up the old cards that we had been playing with. "Think of the new rules as constituting a separate deck of cards," I suggested. "We've seen four of those new cards already. I'll deal them out for you this time. Take another quick look.

"Now, I will deal out four more and explain them as we go along. Oh, and by the way, the symbol on the rule cards—an arrow circling back on itself—signifies that the card can be played repeatedly, and under many different circumstances:

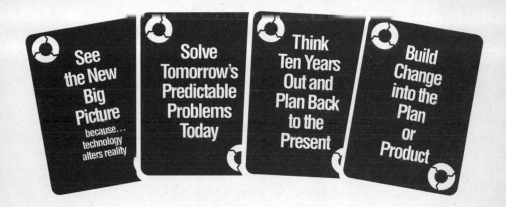

"I don't like to single out any one of the rules as *the* most important, but 'See the New Big Picture' comes close. The first thing to do is to accept that there is a new big picture, even though we might not be able to see it clearly yet. It's out there thanks to the profound impact of emerging technology; we just have to start looking for it.

"We've talked about the new big picture for Blockbuster; let's try another industry. Stein Optical is doing great these days. There are tens of millions of people out there who need eyeglasses or contact lenses, and Stein Optical can serve customers quickly. With many locations they seem like a good bet. But before you rush to invest in Stein

Optical or another optometry chain, look at the new big picture as it is shaped by changes in technology.

"You might remember reading in the late 1980s about a team of Russian surgeons who refined a radial eye surgery technique first pioneered in Japan. They would cut a sunburst pattern of little slits in the cornea of the eye to reshape it and thereby adjust it to the changes brought on by aging and other factors that result in diminished vision." I pointed to Maury and Brenda Jacobs, who were both wearing glasses. "I notice that you two didn't rush out and get little slits cut in your eyes."

"No way. I work in a hospital, and the word was that the Russians hadn't quite mastered the technique," Brenda said. "Besides, I get hysterical at the thought of even putting eyedrops in my eyes."

"Me too," I said. "At first the new technique didn't work all that well; the incisions apparently weakened the structure of the eye. But since the Russians did those early experiments, the excimer, or cold laser, has been developed. There's no scalpel and in the case of using a laser, no heat, which is one of the drawbacks with conventional laser eye surgery. The excimer laser fires off short, high-energy pulses of cool ultraviolet radiation that vaporize microscopically thin sections of the cornea by breaking the chemical bonds between molecules without affecting neighboring tissue. Keep in mind the large number of aging baby boomers. They will want to look natural and young even when they are old. I know I will. I can safely predict they will not like bifocals. I know I won't. By the end of the decade, many who wear eyeglasses or contacts are not going to be or will be seriously considering having this relatively simple, painless procedure performed, which will eliminate the need for corrective lenses."

"You mean I'm going to have to go back to being green-eyed again?" Samantha asked.

"No, contacts that are worn to change eye color or filter the sun's harmful rays and glasses that suit an individual's favorite 'look' will always be around. But in ten years, the majority will be clear lenses with an ultraviolet filter."

"That could be a huge new profit center for hospitals," Brenda said, shaking her head in amazement. "Imagine the number of people who are Stein, Pearle, or Four Eyes customers trooping into hospitals or ophthalmologist's offices to have their vision corrected. There hasn't been anything like it since the nose job."

"But you're assuming that those optometry chains will just roll over and lose the business," Maury said. "If I were running the Stein Optical franchise—which would probably be a lot easier these days then my mainframe computer business—or running one of those other outfits, I'd hire ophthalmologists and set up my own clinics to do the eye surgery."

"Bingo! You see the new big picture," I said. "The change won't happen overnight. There will still be a strong demand for lenses and frames for the rest of the decade. Some people won't want to mess with their eyes under any circumstances. Even so, the sale of lenses and frames will start to flatten out fairly quickly once clinics and hospitals start offering the service. An optician who assumes that tomorrow will be pretty much like today better be only a few years away from retirement; otherwise, he or she is in for tough times."

Tanya started rummaging around in her brown-leather shoulder bag. She held up her sunglasses. "What about these? Couldn't opticians go more heavily into sunglasses both as a fashion statement and as protection once the public realizes the sun's negative effects on their eyes from the reduced ozone layer of the atmosphere?"

"Congratulations! You too are seeing the new big picture," I said. "Simply by recognizing that fundamental change is taking place, you are able to re-examine your basic assumptions. Opticians currently sell 20/20 vision. They tie the price you pay and the value of what they do to the frame or contacts you buy. They compete using price and speed of service. Technology, however, is going to eventually eliminate the high demand for their main products—at least in terms of lenses and frames. To stay in business they'll have to come up with another product." I stopped and reached over the table for Tanya's sunglasses. "Instead of selling frames and competing on price, they could sell the service of fashion coupled with total eye care, including eye protection. Taking the focus off of the price of the frames and putting it on the quality of the service is in keeping with an optician's existing customer base and business experience. Samantha mentioned eye color, which is a logical product focus for soft contact lenses— blue eyes to go with a blue dress. Also, as we age our eye color can fade. Special contact lenses that restore our original eye color would make sense."

It Doesn't Grow on Trees

"Nice glasses, Tanya," I said, trying them on.

"Thanks. It's the one thing I allow myself to splurge on."

As I handed them back across the table, I noticed a bowl of fruit on the sideboard behind her. "I'll trade the glasses for one of those oranges," I said, nodding toward the pewter bowl.

Tanya turned around and selected the largest specimen. "Overhand, underhand, or shot-put?" she asked, hefting the orange as though it were made of solid cast iron.

I accepted the orange from Tanya with both hands and held it up for all to examine. "Here's another example of seeing the new big picture—or failing to see the new big picture. What is this?" I asked, looking at Brenda.

"An orange," she said, with a shrug of her shoulders.

"Oh, really?" I slowly peeled off the waxy orange skin and broke the fruit into segments. Without commenting, I squeezed the juice out of each segment into an empty coffee mug. "What you identified as the orange, Brenda, was really the orange's packaging material." I pointed to the peel and the soggy pulp. I put the cup to my lips and sipped. "The point I'm making is this: Technology is about to alter the reality of oranges and other fruits and vegetables. Scientists have developed a technique that gives them the ability to grow large quantities of any individual cell from a wide variety of plants. There is a biotechnology company in Florida that is growing the inside of an orange—no peel, no pulp, no seeds, no tree—just the cells that produce the juice." Turning to Maury, I said, "You work in a large building downtown, Maury. If that building were used to create orange juice, it would take the place of several thousand acres of citrus groves. Plus, there would be no worry about early frosts, blight, the use of pesticides and herbicides, or droughts."

I paused before I asked the next question. "What's the new big picture for Florida citrus growers?"

"Parking lots," Tanya shot back. "And shopping malls." I had obviously hit a raw nerve.

"No, the future is not that grim for farmers, Tanya. Unless they—and you too, of course—refuse to see the new big picture. Those who insist on doing the same old things in the same old ways will be flipping hamburgers and waiting on tables at a fast-food chain."

"Dan, if I were a citrus grower looking for the new big picture, I'd first try to get a license on that process, since I'm already in the orange business and know the drill," Glenn observed. "Then, I'd form a partnership with that biotechnology firm and use my land, equipment, and labor to experiment with other crops and agricultural techniques. If they weren't interested, I'd go to the nearest university and see if some young Ph.D. candidate could come up with bright ideas and a research grant to turn my citrus grove into a big outdoor laboratory. If you can't lick 'em, join 'em."

Glenn was right. It is futile to fight the new big picture; the goal is to become part of it.

Foresight vs. Hindsight

"We've all become expert crisis managers by default. And that's why the seven of you are here with me tonight," I said, moving on to offer a brief explanation for each of the new rules. "Your businesses and careers have become an infinite series of crises. As a result, you never have time to manage your opportunities. By *solving tomorrow's predictable problems today*—one of the new rules that I just dealt out— we can head off a crisis before it even develops.

"Think about your biggest problem. How many of you had a current problem in mind?" They all nodded yes.

"Now consider your major problems from five years ago. Most of them, looked at with the benefit of hindsight, were probably predictable, and the amount of time and resources spent solving them was totally out of proportion to their future significance."

Doug Stedman nodded enthusiastically. "That happens all the time at school. There's a huge flap about something or other, and six months later nobody can remember what the fuss was all about."

As a former educator, I knew exactly what Doug was getting at. "How well I remember, Doug. The trick is to substitute foresight for hindsight. Most school administrators haven't learned how to do that. I don't mean that it's necessary to become clairvoyant. When do people get burglar alarms?" Doug and Maury quickly responded, "After they get robbed."

"What it takes is the willingness and the wisdom to step out of the present for a moment and consider the future," I said. "With the benefit of foresight, we could have seen the savings-and-loan crisis loom-ing, the credit crunch that hit small businesses in the early 1990s, the cutbacks in defense spending, and the emerging paperwork bottle-neck.

"Would everyone agree to the proposition that what we do today shapes tomorrow?" I didn't hear any objections. "Good. If we spend the present running around putting out fires, our future will be more of the same unless we take the time to stop and buy some fire insur-ance."

"Now you're talking," Samantha quipped. She was a member of the life insurance industry's elite Million Dollar Round Table, and I was certain that before the card games came to an end, she would be fully conversant with each of our insurance needs.

I picked up on Samantha's comment and ran with it. "As Saman-tha can tell you, insurance is a real challenge to sell. Why? Because the customer is being asked to step out of the present and purchase a product that he or she will use in the distant future—money to pay for medical expenses or disability or to support one's family."

"Talk about solving predictable problems," Samantha exclaimed. "How are the kids going to go to college if I die? Life insurance, that's how. You can solve that problem today."

"Death is certainly a predictable problem," Lonnie said, prompting Glenn to add:

"And taxes."

"Thank you Ben Franklin, but stay with taxes a minute," I urged. "Lonnie, if your landscape design firm starts expanding, you are headed for a far more complicated tax situation. Solve that problem today by consulting with a CPA rather than waiting until the week before April 15 when it is too late.

"As for death, the other component of Franklin's famous apho-rism, when do many people start fitness programs? Usually when the doctor says 'You're fifty pounds overweight . . . better start getting some exercise or you'll have a heart attack one of these days.' The obesity problem has a head start. The more sensible thing to do is to anticipate, initiate, and obviate. Instead of waiting for the flab to set in, anticipate the consequences of the metabolic changes and a more sedentary life-style that middle age brings, initiate a solution, and avert, or obviate, the problem before it occurs.

"The best aspect of this kind of forward thinking is that it allows you to become *pre-active* rather than reactive. This is the essential characteristic that differentiates the crisis managers from the opportunity managers."

Sweet Serious Dreams

Glenn pointed to another of the cards that was lying faceup on the table. He read the rule aloud: "*Think ten years out and plan back to the present.* That doesn't seem possible if change is coming on as fast as you say it is. Can a ten-year plan work under those circumstances?"

"Yes, it can if you observe the next rule," I said. "*Build change into the plan or product.* The two rules complement each other. By going ten years into the future, you are forcing yourself to escape from the tyranny of the present. To set goals that far in advance, it's necessary to do some serious dreaming. And that's the secret of successful long-term planning. Come up with a wish list of things that you would *ideally* like to be doing or goals the company should be pursuing. Don't be inhibited; really stretch.

"Most plans fail because they start with the present—year one—and proceed to year two, year three, and so on. As a result, all the specifics tend to be front-loaded and based on current reality. As the plan carries forward it gets more and more vague. By planning backward, from the future to the present, I am better able to see where, when, and how to take action in order to get to my ideal objectives."

I stopped talking long enough to find a photo album that I keep in my bookcase. There are three or four pages of pictures recording a sabbatical I took over a decade ago mountain climbing in Glacier National Park in Montana, a goal that I had set to mark my thirtieth birthday. "This shot was taken when I was heading out from Swiftcurrent Lake for an all-day hike up Altyn Peak," I said, turning the page and holding up the album for everyone to see. "As I got to the base of the mountain, I worked out what I thought was the best route to the top by simply looking at the summit and saying 'There's where I want to go.' I traced a route down the mountain to where I was standing. About an hour into the climb, I cleared a ridge line that had been blocking my view and realized from my new perspective that there was a better path to the top. About another hour later, after negotiating a series of switchbacks, I saw a third route that was even more direct.

It was a good thing that my predetermined plan was not too rigid. As a matter of fact, from where I was standing on the trail, I could now see that what I had originally thought was the summit was an outcropping that was about a quarter of a mile below the real peak of the mountain. Even my goal turned out to be off the mark and had to be changed. I wanted to get to the top, not to what I thought was the top from my view at the bottom. I can remember thinking that I would have never known my true destination if I hadn't started the journey."

I closed the album. "The point of the story is our new rule: *Build change into the plan or product*. It's just as important as having a plan, especially in times of rapid change. Periodically, review the plan and adjust it to fit changing circumstances. You've got to be flexible. The ultimate goal may not change, but the route probably will.

"As for building change into a product, think of the Touch-Tone phone. When it was first introduced people asked, 'What are the two buttons next to the zero for?' The answer was 'nothing.' The designer put them there, figuring that someday there might be a use for them. And that's one of the reasons the Touch-Tone phone is still in use.

"The early Apple Macintosh computers, on the other hand, didn't have change built in. The machines were touted as all the computer the consumer would ever need and were manufactured with no expansion slots. Before the end of the first year of sales, many customers had removed the back of the computer and added third-party components, even though it meant invalidating their warranties. Eventually, Apple realized the mistake, and the company added an expansion slot to allow customers to make changes quickly and easily."

Lonnie picked up one of the previous cards and read the rule: "*If it works, it's obsolete.*" He flipped it back onto the table. "It sounds like Apple learned two of the rules the hard way. By *building change into the product or the plan* you are also avoiding obsolescence. It's always a work in progress, and the moment you write *the end, finis, complete* that's when the dust begins to accumulate."

I was delighted to hear the comment. It meant that the rules were beginning to sink in with my card players. The object of the game is to transform the way the players think, not to impose a set of hard-and-fast edicts that are memorized but never followed, accepted, or acted upon.

"Let's look at four more rules," I suggested.

"These cards, as you can see, are customer oriented. It's funny, a lot of people think that they are in jobs and careers that have nothing whatever to do with such a grubby and messy thing as a *C-U-S-T-O-M-E-R*. But I don't care who you are or what you do, there is a customer out there making yes or no decisions that vitally affect your most basic interests."

Doug Stedman cleared his throat and spoke up. "I'm probably the one exception, Dan. In fact it's one of the best things about teaching—no customer to deal with."

Tanya quickly joined in. "That's what keeps me in farming," she said. "I'd go crazy if I had to deal with customers."

I could see Maury Andrews's jaw drop. "You buy that, Maury?" I asked, suspecting what the answer would be.

"Buy it? Not at any price! Kids and their parents are a teacher's customers. If teachers treated them as customers who had a choice, maybe we wouldn't be in the middle of an educational crisis in this country," he said, looking straight at Doug.

His friend and fellow CEO had been right. Maury could certainly be blunt. Doug responded, "I could defend myself, but I've got to admit you've made a good point. I was never taught about customer service probably because our customers couldn't leave. They have to stay until they are at least sixteen."

I jumped back in to head off further dispute. "Tanya, there is a huge milk surplus in the United States. Why is that?"

"People don't drink enough milk," she replied.

"Reverse that equation," I suggested. "Consumers—customers—

are buying and drinking all the milk they want. But beyond the volume that meets current demand, farmers are producing a product that the customer has no need for, and yet they continue to do it anyway. If Maury or Glenn, Lonnie or Samantha—even Brenda's hospital—did that, they'd be out of business in short order. In fact, we could make the case that Maury and Glenn, all of you for that matter, are going out of business in slow motion precisely for that same reason." I figured I might as well get everybody riled up.

"Now wait just—" Maury interjected.

"Okay, start with you, Maury," I said, cutting him off. "What you said about kids and their parents being a teacher's customers was right on the mark. But when was the last time you dealt directly with one of your company's customers?" The question stopped Maury in his tracks.

He thought for nearly a full minute before answering. "The beginning of last month, but that's not important, since my people are in regular contact."

"What people?"

"Sales reps, district managers."

"That's why God invented salespeople—to deal with customers," I said, smiling at Maury. "Aside from private dining rooms and fancy bathrooms, one of the perquisites of getting to the top of the corporate hierarchy is not having to deal with customers. Many times, as executives climb the career ladder they have less and less contact with the customer. Eventually, anything but the most perfunctory exposure is actually considered to be demeaning.

"You jumped Doug for not knowing who his customers really are, but at least he deals with them all the time. Doug's bosses, however, are a different matter. As many school administrators advance in their careers, they lose touch with students and parents. Naturally, when it comes time to design a curriculum, they begin to founder."

Doug saw his opening. "They are still teaching woodworking in my school," he said. "Not too many years ago, it would have prepared a student for the future. Just like the rule says: *Focus on your customer's future needs.* But there's no real cabinetmaking industry in our area. Even if there were, the fabrication techniques and equipment would be totally different from the chisels, vises, and handsaws the kids are taught to use."

I seconded Doug's point. "There's nothing wrong in learning

woodworking for making home repairs, as an art form, or as historical hands-on learning. However, we have to look at the emphasis we are placing on any subject and ask, are we educating those students for a day gone by? A thirteen-week summer vacation to allow children to help their parents in the fields is another example of yesterday thinking rather than future thinking.

"By not focusing on the customer's future needs, an education system or a business dooms itself—and the customer in many cases—to failure. Those future needs are not going to be the same as current or past needs, thanks to technological changes. By looking at the future through the customer's eyes, we can see our own future more clearly."

A Double Play

Brenda Jacobs raised her hand for recognition. "I suppose schools and hospitals have a lot in common. We do things at the hospital to suit our own convenience, which is not necessarily convenient for the patient. In some ways, we are obsessed with ourselves organizationally as opposed to paying attention to what the patients need."

"Brenda, you've just said it all," Doug replied. "Those woodworking courses are still being taught because it is too much trouble to drop them—trouble and threatening for the shop teacher, trouble for the principal, trouble for the school administration. They have a fundamental investment in the equipment, and in this case the past."

"Let's finish up with this rule," I recommended. "Simply stated, *focusing on your customer's future needs* is a way to focus on your own future needs. In 1992, Philips, the Dutch consumer electronics manufacturer, introduced a digital tape recorder that uses both the old analog cassettes and new digital CCD format cassettes with near CD-quality sound. It's a product that focuses on the future need for a superior audio medium even though most customers possess a large number of the old-style cassettes. If Philips was self-obsessed, to use Brenda's term, it would continue to build a standard cassette machine because it already has the know-how and manufacturing capability."

Tanya brought the discussion of the rule to a close by observing that Philips was also using an earlier rule: *Build change into your plan or product.*

Having It Both Ways

I held up the card for the next rule. "*Sell the future benefit of what you do*. Here again, there is future benefit built into the Philips tape deck. Whoever sold the company on the idea was in the position to offer twice as much product for the money. Unlike the Sony DAT digital recording decks that were introduced a year earlier, it meets a current need and a future need in one package."

"It is also a way around the obsolescence problem," Glenn remarked. *"If it works, it's obsolete."*

I nodded in agreement. "Not only that, it amounts to hedging a bet. If it takes longer than anticipated for the public to embrace digital tape-recording technology, Philips still has analog to fall back on. Plus, it will capture the market for digital tape-recording."

"Dan, earlier you said that as an insurance agent, Samantha has to surmount a difficult challenge because she is selling the future benefit of life or disability insurance. Isn't this new rule going to create a similar problem?" Brenda asked.

"Not if it is sold as an extra dimension," I said, and Samantha also had a reply to the question.

"I try to have it both ways. Dan was right. People really don't like to think about the future, especially if it involves sickness or death. What I do is tell my youngest customers, the ones who don't have any credit rating, to use me as a reference. After all, they are making regular payments and demonstrating their reliability."

I was impressed. "Pretty slick. Samantha is selling credit ratings—a present need—and throwing in, as an extra, the future need for money to pay doctors' bills."

"It's a different kind of thing, but when we had coffee, Glenn and I were talking about all the frequent flyer mileage we've accumulated," Lonnie said. "The airlines are using the rule. That's a present need—transportation today—and the future benefit of a free trip next year."

Maury and I started to respond to Lonnie at the same time, and I yielded to the CEO. "That's definitely a future benefit," he said. "But the airlines are giving away their future to stay solvent today. If it gets out of control and goes beyond just using excess seating capacity, those frequent flyer programs are a crutch that could end up crippling the airlines."

I nodded in agreement. "That's an example of solving today's problems by mortgaging the future. It totally overlooks the new big picture."

A Better Path

"You seem to be getting the hang of this," I said. "Any thoughts on the next rule, *Build a better path to the customer?*" There were no takers, so I continued. "Ideally, technology determines the path to the customer—unless we get stuck in a rut." I noticed that Samantha had added this last point to her notes and drawn a large circle around it, so I decided to reinforce the concept on my flip chart.

Technology Determines the Path to the Customer

"The traditional path to a grocery store was through the front door of the neighborhood market," I said. "Walking or riding a bus or trolley to work was the traditional mode of transportation. When we started to change our living patterns, spending more and more time behind the wheel of a car, the grocery chain built larger stores with lots of parking in the suburbs and pretty much gave up on the corner store." I pointed to the front windows where we could see Maury's limo at the curb. "This was all in reaction to technological change.

"Today, transportation is starting to bog down. A trip to the grocery store on Saturday can turn into a major hassle. Yet, most grocery chains are still using the old pathway." I looked over at Glenn as I continued. "You're a food distributor. If you could help your grocery stores build a new and better pathway to the customer using television sets, or PCs, Touch-Tone phones, and home delivery, wouldn't that benefit your bottom line?"

"Sure, if it would increase sales volume."

"Volume is only one aspect. If the retail customer is doing his or her shopping from home without visiting the store, there'd be less of a need for you to spend time and money restocking the retailer's inventory. In fact, for a fee you could do the deliveries from your own warehouse."

Glenn asked, "Are you saying that stores and shopping malls will become obsolete?"

"No! There are many items that we will not buy if we can't touch, kick, try on, sit in, or smell them. Home shopping is best suited for items that you don't need to physically interact with in order to buy, like a can of beans." Glenn immediately started doing mental arithmetic on the costs.

"We were talking before about airlines and frequent flyer mileage. Why not go down to the bank, or the grocery store for that matter, and get your tickets out of the automatic teller machine? The ATM gives the airlines a better path to the customer."

Could've, Should've, Would've

"How about this rule: *Give your customers the ability to do what they can't do, but would have wanted to do, if they only knew they could have done it?*"

Doug spoke up immediately. "ATMs did that. After 2:00 P.M., when the banks in my area closed, you were out of luck if you needed cash, particularly on a weekend or if the liquor and grocery stores were closed."

"And forget it if you were out of town," I added. "But we didn't even know there was another way to get cash until the ATMs arrived on the scene. A friend of mine who is a real Luddite at heart insists that ATMs are the only truly useful technological development since the invention of the electric light bulb.

"Technophobes like him love to use the videophone as an example of a long heralded breakthrough that never amounted to much. But the videophone's problem is that by just showing us a picture of the other party, it is giving us the capability of doing something we don't really want to do for the most part."

"Amen to that," Samantha said. "If I had to worry about what I looked like every time I made a phone call, I'd never get any work done." The rest of the group agreed.

"Conversely, new technology fails when it does what we are already doing just as well with the old technology. A computer program that keeps our checkbook straight and lets us know when we are overdrawn is nice, but the real problem is having the discipline to log the information and keeping enough cash in the account to begin with."

"Dan, do you subscribe to CompuServe?" Lonnie asked. I said that I did. "It's got a shopping service that proves your rule. I've never used it. I can flip through a catalog or tune in a home shopping network on my home television and not have to bother with passwords, and modems, and that kind of foolishness."

I agreed and added, "Most people—the ones who aren't computer hackers—feel the same way. But when it comes to things like electronic bulletin boards and specialty forums for exchanging information, CompuServe and Prodigy and other services like them are providing consumers with a product they can't get anywhere else. With a much better user-friendly interface using advanced 2-D simulations, the use of electronic services will really take off."

It was starting to get late and I wanted to leave the group with a few more rules. I dealt three cards faceup onto the table.

Time Is the Currency of the '90s

Leverage Time with Technology

Enter the Communication Age

"How often have you heard, 'So much to do, so little time to do it'?" I asked.

There was laughter, and Brenda replied, "How often have we heard it? How often have we all said those words is more like it."

"Back in the 1950s," I continued, "technology was sold as a miracle that would allow us to work less and play more. Futurists were predicting that filling up all our leisure time was going to be a major worry. But what happened? We found more to do. We created new products, new markets, and new global customers. In addition, many large companies cut personnel to lower costs, which left fewer people to do the same amount of work. This has given many of us less leisure time. Unfortunately, we often use technology to do what we were already doing, only we now do it faster. What irony! The future turns out to be the same old thing, just cranked up to a zillion miles an hour.

"Therefore, to many of us, time is more precious than money. I don't know about you, but every now and then I get a slow pump in a self-service gas station. I hate slow pumps because my time is too valuable to watch the gas slowly trickle out of the nozzle. I will put a few dollars worth of gas in the tank and then I'm out of there. This is, of course, stupid of me because I only have a few gallons of gas in my car. I will have to pull into another gas station soon, and even if it has fast pumps, it will take more of my valuable time than if I would have filled it up in the first place with the slow pump. Do you think I will ever be back to the station with the slow pump? No! Even if they put up a sign saying 'We have fast pumps now.' No! Why? Because I still perceive that my time is too valuable to waste on a slow pump, and today we all have options. There used to be one convenient location for a product or service that we wanted, now there are many. The old rule of customer service was, three strikes and you're out. The new rule is, *one strike and you're out*. Remember, if you give me time, or sell it to me, I have received something that has intrinsic value. Hence, the rule: *Time is the currency of the '90s*."

I turned to Doug. "I was critical of long summer vacations for students and teachers, but I should ask you if you would willingly give up those vacations?"

Doug looked horrified. "That summer break is worth $10,000 or $20,000 as far as I'm concerned. I would rather have the time than the money."

"That's what I'm getting at," I said. "The next rule, *Leverage time with technology,* is the way to turn time *into* money. It doesn't have to be an either/or situation.

"Here's a favorite story that I have told countless times on the lecture circuit. It involves a visit I made to a brake and muffler shop beside a shopping mall near my home. After filling out the paperwork at the counter, I asked the manager when my car would be ready. He told me to take a seat in the waiting room and they'd let me know.

"'You should use pagers,' I said.

"'Pagers?' the manager replied, giving me a sideways look. 'We don't even have a PA system. This place is small enough that all I do is holler if I need somebody.'

"'No, I mean give pagers to your customers so that they could go over to the large mall next to you and do some shopping while the car is being worked on. You could page them when it was ready. I'll bet you will have more customers.'

"The suspicious look the guy had been giving me changed to one that seemed to say 'Hey, this fella with the beard isn't so weird after all.'"

"That's terrific," Brenda said. "In our outpatient clinics and labs we keep people waiting so long that there's nothing but complaints. We could hand out pagers, and the patients could go to the snack bar or across the street to the bookstore to kill time."

"The bookstore might consider going fifty–fifty on the cost of the pagers," Samantha suggested.

The ideas were sparking other ideas for Brenda. She plunged ahead saying, "Actually, every ambulatory patient should get a pager. We lose track of them when they start moving around for laboratory tests or therapy. All we'd have to do is buzz the pager, and they could go to a house phone to report in."

"I think I'll start a business supplying pagers to hospitals," Lonnie said.

"Okay, before we get carried away with entrepreneurial enthusiasm, let's do one more rule," I urged. *"Enter the Communication Age."*

"Haven't Alvin Toffler and John Naisbitt and other business gurus said that we are in the Information Age and that by the year 2000, the Information Age will be in full bloom?" Maury said.

"I don't want to be in the Information Age," I answered. "Information is static; it just sits there taking up space. Communication is

dynamic and causes action. Computers and telecommunications technologies are merging to give us the ability to create the Communication Age. Don't confuse information with communication. Technology
has given us the ability to accumulate vast quantities of information to
the extent that we are now unable to digest it all." I stood up and
gathered the cards together to signal that the evening was ending.

"When we get together tomorrow night for our next game, I'm
going to give you a simple and vivid example of how most of us are
being tyrannized by the Information Age and a simple way to set ourselves free."

3

Playing with a Full Deck

New Cards for a New Game

When we assembled the next night to continue the game, both Maury and Lonnie had a pair of eights, beating all the other hands around the table. Lonnie reached for the pot.

"Slow down," I said. "If we were playing by the old rules, you'd win even though your pairs are the same value."

Lonnie nodded. "Right. Out of the remaining nonpair cards, my highest is a seven and Maury's is a five. I win." He went for the pot again.

"No, Maury wins because by now all of us here have seen the new rule—*Build a better path to the customer*—and that gives him an advantage that the rest of you can't top."

There was a chorus of protests from around the table. "If we all know and follow the rule," Samantha Brewer objected, "how come he has the edge?"

"Maury's company is huge," I explained. "Lonnie can build a better path to his customer, but that customer base is tiny in comparison. If Maury is using the rule, he's got far superior leverage and resources that can't be matched."

The CEO smiled and raked in the chips from the middle of the table. Unlike his stubborn insistence on playing with the old rules at our first game, Maury was making a real effort to use the new rules. His game had improved, but he was still hampered by his relative lack of knowledge of the new technology that had changed the rules in the first place. The other players were having similar problems.

I was planning to introduce them to the new technological tools before the evening was over. In the meantime, though, I wanted to make sure that everyone understood that the new rules are not just for start-up companies, entrepreneurs, or mom-and-pop operations. Major companies like the one Maury runs are not automatically fated to go the way of the dinosaurs if their leaders are willing to accept the new reality and take appropriate action.

"But if past success is Maury's worst enemy," Samantha persisted, "and since he is more successful than the rest of us, isn't he at a permanent disadvantage?"

I had been pleased with the way Maury had been discarding the high-value face cards during the first few hands of draw poker that we had played. He was definitely trying to see the new big picture. "What about it, Maury?" I asked, giving him a chance to respond to Samantha.

"I've been thinking a lot about this since our last get-together. At first, what you're driving at, Samantha, seems only logical. But when I took a second look at some of the other new rules, like *Make rapid change your best friend, Learn to fail fast,* and *Build a better path to the customer,* it occurred to me that while a struggling little company can do those things, I could probably do them better and faster, considering the size of my corporation. We've got a line of credit at the bank that would stretch from here to the moon and back. I could finance a string of expensive failures and none of it would show on my bottom line for years." He turned to Lonnie. "If you tried to do that with your landscaping firm, the Chapter 11 boys would be pruning your rosebushes."

"With a chain saw, no doubt," Lonnie said.

Maury was idly arranging his chips in five neat stacks. "Name recognition is another strong point that my company has going in its favor. I don't have to build a reputation transaction by transaction. If I announced tomorrow that we were going to use the new cold laser to correct vision problems, people would sit up and take notice."

"You sound like a convert to me," Doug Stedman suggested.

"I don't know about that. I am still opposed to change for the sake of change. A few years ago, we tried to flatten our corporate organizational chart. We set up several autonomous divisions. Now, the left hand doesn't know what the right hand is doing. It was probably a mistake. If I am becoming a convert, it's because of Dan's comment about going out of business in slow motion. Big companies are particularly vulnerable to that kind of thing. The signs of death are hard to spot from thirty thousand feet in a private corporate jet. At least you're showing signs of life and putting up a fight when you embrace change."

"And you are putting up a fight when there's still enough strength left to do battle," I said. "The Japanese did U.S. business a favor by coming on as strong as they did in the 1980s. If they had continued to tiptoe around those slumbering giants, many of them would have died in their sleep."

"Detroit got one hell of a wake-up call," Brenda Jacobs observed.

Cleaning the Attic

"Before we go any further, I want to complete the unfinished business from our last session," I said, moving to the flip chart and turning the page.

Enter the Communication Age

"As I told you last time, for about a decade we have been in the Information Age. Now, there's been what I call a 'shift in focus.' It is vital that the Information Age must now give way to the Communication Age." I looked at Glenn Clayton before asking the next question. "Do you still have your college notes?"

"You bet," Glenn replied. "They're in six boxes up in the attic. Every time there's spring cleaning my wife tries to throw them out, and I cart 'em back up there."

"Anybody else have their college notes?" I asked the group. Everyone except Tanya raised their hands.

"I've got notes back through my freshman year in high school," Samantha said.

I wasn't the least bit surprised. "I guess you're all afraid that if you lose your notes, you'll lose your minds." The wisecrack drew a healthy chuckle.

"Glenn, your wife was right. Get rid of those six boxes. You're waiting until you're pushing ninety-five, about to die, and you're going to say, 'Wait, I gotta see my notes. I've been saving them all this time.'

"The same goes for everybody here. You haven't looked at those notes in years. There they sit in the attic or stuffed in some filing cabinet, absolutely useless."

I turned back to Maury. "You mentioned setting up several autonomous divisions that fail to communicate effectively with each other. I'll bet those units are awash in information, but like your college notes, much of the material is a complete waste of space. In the Communication Age we must learn to extract the knowledge from the information, put it into a dynamic form, and communicate it to cause action. By dynamic information, I mean digital information that is focused and linked to other digital information. That way the information that I need can find me when I need it.

"I'll give you an example of *failing* to be in the Communication Age. A few years ago, a major midwestern corporation invited me to speak at an important conference. Two weeks later, I got a request from the same company for information about myself, my program, and my availability. When I called to find out what was going on, the meeting planner I talked with—he worked at the very next desk to the executive who had just hired me—didn't know I had already been working with his company.

"The company had not yet entered the Communication Age. It had plenty of information—often useless information gathered and stored at great expense. Two men, sitting five feet from each other, were totally out of touch, yet each had plenty of information.

"There's no excuse for it. Computer software can link databases and files to automatically update any reference to an upcoming event. Whatever record was available showing that I was planning to speak could have—should have—posted that new information in every other appropriate context.

"Or, if the company used its desktop workstations to dial outgoing telephone calls, the system could be programmed to recognize 'Daniel Burrus' when the name was logged in and ask the meeting planner if he would like to see a menu summarizing all contacts with Daniel Burrus within the last sixty days. A quick check would have avoided the duplicated effort and confusion."

I offered them another example: "All of us go to business meetings and annual professional conferences—and we hear some interesting things. What happens? Out come the notebook and pen. What happens to our notes when we get home? Those notes go into your 'important notes folder,' the same folder you haven't looked at since the previous year's meeting. It's the college notes syndrome under another guise, and that's absurd.

"What I want you to do from here on, whenever you take notes, is to condense your notes into key points—the hot buttons that hit you— when you are on your way home. Put your key points into your computer system and link them to topics that they relate to. You will now have intelligent, focused notes that automatically access themselves when you need them. The beauty of it is that remembering where you put them is no longer necessary. The notes will find you when the subject comes up, and they'll appear in a little window. That isn't fancy programming. You can do that with your hardware right now."

Lonnie raised his hand. "I pick up all kinds of tidbits about certain plants; oftentimes it's just practical experience about what works and what doesn't. But then I'll forget all about them. With what you're talking about, I can program the computer to respond with watering instructions for azaleas every time the subject of azaleas comes up."

"That's right," I said. "It will keep reminding you until you remember the information on your own and deliberately remove the message from your system. The strategy is to *update and then eliminate or you will accumulate*."

"I don't think my hospital's computer system has that feature," Brenda said.

"If you are using current versions of your software, the chances are excellent that you can link the information as I described. What's happening most of the time is that we go to the trouble of buying cutting-edge technology and then use it to do what we did before except a little faster. That's what your competition is doing. All you are doing is keeping up. If you are paying cutting-edge prices for cutting-edge technology, you need cutting-edge applications to gain a true advantage. Talk with your hospital's computer system programmers. I'll bet they can set up a series of what are known as macros that will display reminders and prompts on the screen whenever certain key words and commands are typed in or accessed.

"Why use just a small fraction of your system's capacity? A lot of money was invested to provide a wide variety of features. It's silly to use it only as a glorified typewriter or adding machine."

"Many of my people say they just don't have the time," Brenda countered. "They're too busy attending to patients."

"Ask them to try spending a half hour to an hour a week. Don't try to do it all at once. It's amazing how many professionals, at all levels in an organization, are using the computer as a fancy typewriter. They are living in the past when the future is sitting right on their desk. It is one of the main reasons that people develop a fear of technology. 'The darned thing doesn't work as well as my old Underwood' is the lament. But we haven't taken the time to learn how the 'darned thing' works. The technology gets used to do what we used to do with the old technology, and pretty soon the boss is wondering why he paid out so much money for such a small return.

"This kind of technological sabotage happens again and again. Many perfectly adequate voice-mail and E-mail systems get torn out and discarded because no one took the time to learn the best way to use them. Think of it as a variation on one of the new rules: *Give your customers the ability to do what they can't do but would have wanted to do, if they only knew they could have done it.* In this case, you are the customer. Look for ways that the new technology can break through and perform a task that you'd want to have done, if only you knew it could be done."

Lost and Found

"By entering the Communication Age we can dig out from under the mass of static (paper-based) information that is being generated by new technology. Are you ready for this statistic?" I paused to make certain the group was fully focused on what I was about to say.

"In 1971, a typical state-of-the-art computer microprocessor handled 40,000 instructions per second. By 1991, twenty years later, the rate was up to 40 million. That explosion in processing power acted like a nuclear chain reaction, and it sent up a vast mushroom-shaped cloud of information. It's my contention that we are still dealing with the fallout using the methods and mind-set that we had in the past. No wonder we feel overwhelmed!

"Saving those college notes in the attic might have made sense in 1951 or 1971. But as a meaningful information storage and retrieval system in the 1990s, it just doesn't work anymore."

"I guess I know how I'm going to be spending the weekend," Glenn said, "shuttling between the attic and the dump."

"That's a start, although you may want to take it to a paper-recycling center," I responded. "The next project is to go into the office and look for the business equivalent of your college notes. Declare war on static information. If it doesn't find you, how are you going to find it?"

"Obviously, you're not," Glenn said. "But I've got tons of paper files filled with important information about customers and products. I just can't bear the thought of chucking it all out."

"There is an important distinction to be made," I said. "We have a traditional respect for the thing called 'information.' That's one of the reasons we collect so much of it and hate to part with it. The old aphorism tells us that 'Information is power.'

"Do you know why? In the past, our relatively limited quantities of information could readily be turned into knowledge using good old brain power. Once it was knowledge, action was then possible.

"Action." I repeated the word for emphasis.

"And there's the distinction: *Information is power only if you can take action with it*. Then, and only then, does it represent knowledge and, consequently, power.

"If you feel there is important information filed away, it would be worth while to convert it to a more accessible, usable form by converting it to dynamic (digital) information using hand-held or flatbed scanners, depending on how much information you have to convert."

"What you're saying, Dan, it seems to me is that what Glenn has in his files is so much wastepaper, since it is inaccessible, although some of it may be valuable. . . ." He broke off, unable to find the right word. "What can I use if information is the wrong term?" Doug asked.

"Protoknowledge," Brenda suggested, touching off a round of good-natured groans from the others.

"You and I must be techno-nuts, Brenda, but I like protoknowledge. In ancient Greek *proto* means 'coming ahead or before.' But maybe we should use the analogy of a seed pearl, which starts as nothing more than a grain of sand. Glenn has got to find a way to get those grains of sand out of his files and into the oyster—his wholesale food distribution business—in order to make pearls."

"How?" Glenn asked.

"There's the question I've been waiting for," I said. "It leads me into a discussion of the next set of cards that we are going to be using—the new tools of technology.

"With the new rules you've been learning a new way to think. The new tools provide a powerful way to shape the future. Incidentally, I haven't shown you all the new rules yet. There are other very important rules that I will introduce as we go along."

"Great, just as I think I'm on a roll and reach for my next jackpot, you're going to spring another new rule," Maury grumbled, although he was quick to follow up with a wry smile.

"Would I do that, Maury?" There was general agreement from the group that I could and certainly would.

The New Tools

"I am going to give you a head start on the twenty-first century by sharing a closely held piece of information—make that 'knowledge' now that we know the difference—and insight. Namely, that when it comes to technology, the next ten years have already been invented; you just can't buy the products off the shelf yet.

"In the early 1980s, I realized that the pace of change would increasingly make it difficult for organizations to anticipate the future and be correct. When change was slower and the world less complex, it was far easier to anticipate changes to gain an advantage. Even less than ten or fifteen years ago, it seemed almost impossible. I knew that if I could develop a method that would allow me to predict the future and be right—and since I'm a professional technology forecaster and futurist, that's the name of my game—I knew I would have something on which to base a highly successful business.

"The first key to being right about the future is to leave out the parts that you can be wrong about. The question then was, is there a second key, the key to anticipating with accuracy?

"History reveals that innovations in technology have always provided a window to the future. I started a research company dedicated to uncovering global innovations in all areas of science and technology. After years of research, I identified twenty core technologies out of which future products and services will emerge. Not two hundred or two thousand. It's a very manageable number—now. It wasn't at the beginning, not by a long shot. There were thousands of interesting and potentially significant technological developments. Many had to be translated, and I had to separate the wheat from the chaff. Where to start?

"Thanks to my background as a former science instructor, I began grouping or pigeonholing them by their primary technological characteristics. Lasers went in one pile, fiber optics into another, and so on. To qualify as a core technology, the common denominator was that each had to be emerging or about to emerge from the development stage of its evolution and enter the application stage with revolutionary impact on a wide range of areas. It isn't the invention date that is important, it's the application date, the date that you and I can put the cards in our hand and play the game.

"What took shape was a highly focused, sharply delineated constellation of less than two dozen technologies that fairly pulsated with energy and potential.

"Knowing what happens in the next year or the next decade doesn't have to be a matter of guesswork and crystal ball gazing. It amounts to utilizing what I came to call the new taxonomy of high technology.

"Don't let the terminology throw you. I'll elaborate on what I mean by *taxonomy*. Years ago, scientists were faced with the problem of trying to make sense of the enormous number and variety of plants and animals around the world. The volume of data, even by eighteenth- and nineteenth-century standards, was overwhelming. The solution, thanks to Swedish botanist Carl von Linné, (also known as Linnaeus), was to simplify the information by organizing it into a classification system, or taxonomy.

"It was a remarkable achievement. Today, thanks to the taxonomy we learned when we were in grade school, we can see an animal we have never seen before and determine whether it will bear its young alive or lay eggs. We are also able to predict with accuracy that certain types of trees will shed their leaves in the fall and others will remain green. Categories emerge—mammal, reptile, deciduous, conifer—and tens of thousands of individual species and plant and animal types begin sorting themselves out; in the process, highly useful generalizations can be made.

"My new taxonomy of high technology—tech(s)onomy?—works the same way by grouping together significant technological breakthroughs in rough subsets according to the obvious specific principles and properties that are involved. I was trying to create chunks of technology rather than slicing it thinner and thinner." I went to the flip chart and turned the page. "Here's what I got."

The New Taxonomy of High Technology

Genetic engineering	Distributed computing
Advanced biochemistry	Advanced computers
Lasers	Artificial intelligence
Fiber optics	Optical data storage
Microwaves	Digital electronics
Superconductors	Advanced video displays
High-tech ceramics	Micromechanics
New polymers	Photovoltaic cells
Thin-film deposition	Molecular designing
Fiber-reinforced composites	Advanced satellites

There was dead silence.

"Don't panic! I don't want us to spend too much time on my taxonomy. We are going to play with the new tools that have sprung from this list. Remember, they represent broad categories, core technologies. Their biggest value is to use them as a window to the future. The most revolutionary new products and services will grow out of them.

"Take any of the chunks, or categories. Genetic engineering is an example. As an item in the taxonomy, it is a synthesis of hundreds of exciting and extremely technical processes, and all of them, as I said a moment ago, have entered the application stage.

"I know, the word *taxonomy* is a daunting word. Most of us probably didn't start hearing it until the eighth grade, and promptly tuned out the teacher and tuned into the joys of the opposite sex. The best way that I have found to approach this complex subject is to use the metaphor of a deck of cards as we explore the opportunities that have come from the list. This linguistic translation makes the technical jargon more user-friendly. Therefore, let's not get lost in the details of the taxonomy.* We are here to learn how to play this new game.

"There are twenty-four of these new tool cards; when nobody was looking they were slipped into the old deck, and that's why the rules of the game have changed." I turned to Doug. "Now, you're beginning to see the cause and effect you wanted. The future will never be the same again—that's quite an effect.

"Remember what I said earlier. You don't need to know the physics of a telephone in order to use it. You have to know that it exists, and you have to creatively apply it to what you are trying to do. Until we learn what new cards exist and what to do with them, we aren't playing with a full deck."

"I am beginning to see the cause and effect, but why the discrepancy in the numbers?" Doug asked. "You said there are twenty core technologies and twenty-four new tool cards."

"Basically, in that taxonomy of mine, there is still too much cause and not enough effect." I tapped the chart. "When this was unveiled, I could see visions of high school chemistry and physics hell flashing through your eyes. You all had the same stupefied look. To remedy that, the cards that we will be playing with each represent a new tool that was made possible by one or more of the core technologies. Unlike the core technologies, they have a heavier emphasis on the practical and most significant applications of the core technology—the bottom line—which amounts to the most important characteristic of all.† The cards are in the deck because they are playable right now. The essential questions are, how are these new tools of technology going to be used to achieve solutions, to remove obstacles, to meet challenges?

"You wouldn't want to be playing a high-stakes card game and not know what all of the cards are, would you? So let's take a quick look at the entire deck." I squared the edges and fanned out the deck.

"Pick a card, Tanya," I requested. She gingerly pulled one from the center of the fan, took a quick look, and turned it facedown. I laughed. "You're going to have to show it to me; I'm a futurist, not a magician." Tanya grinned and turned the card up.

*See Appexdix A.
†See Appendix B.

"First, let me explain the symbol on the cards," I said. "All the tool cards bear the infinity symbol to show that they can be used in an infinite number of ways and in infinite combinations. The only limitation to how we might apply any of the tool cards is our imagination. Now, the card Tanya drew, EDI, as it's commonly referred to, is a good place to start. As good as any. By drawing the cards at random, I'm demonstrating that there is no predetermined order, or technological hierarchy. Each card is powerful in its own right and is independent of the others. At the same time, though, the cards all interact. By mixing and matching various cards, the player can put together powerful combinations. What's more, the new rule cards also interact with the tool cards and form potent winning combinations."

"Dan, are all the cards part of the same deck?" Maury asked.

"Yes, they are. But if it makes it easier to visualize, think of it in terms of a draw poker game using two decks. One deck is the new tools, and the other the new rules. It is important to keep in mind that you can still use the old cards, the old technologies, when playing the game. The new cards have basically allowed you to change the rules and gain an upper hand. Most of the cards that you will find yourself discarding from your hand are the old cards that no longer apply, and you will be replacing them from the new rule and tool decks. Since we are focusing on learning how to create new winning hands, we will leave the old tool and rule cards out of the game; after all, you already know how to play them.

"For now, I'm only going to provide a brief description of what each tool card is and how it might be applied. That's all you'll need. You'll learn how to play the game after we take a quick look at each card. After all, you don't want to be playing with a half deck, do you?

Let's use the EDI card Tanya turned up as an example. This is probably best described as a 'club sandwich' technology. It is several layers of scientific and engineering breakthroughs piled on top of one another. From our taxonomy of core technologies it utilizes distributed computing, lasers, advanced computers, digital electronics, and in some cases, fiber optics and microwaves. This tool is a good example of how the technologies interact.

"Now let's look at the application. EDI has the power to revolutionize the delivery system of products and services. This card will forever alter relationships by allowing companies and industries to leap the boundaries imposed by traditional methods of communication—paper and voice—and automatically link up their business transactions in a way that will give both large and small businesses the ability to lower inventory costs and increase customer service. Sprawling multinationals will have the intimacy, control, and quick turnaround time of cottage industries. For example, EDI lets the cotton thread manufacturer in Taiwan know that two more spools will be needed by the denim mill in Georgia because The Gap in Washington, DC, has just sold three pair of jeans."

I asked Tanya to select another card.

"This new tool seems like an old tool. But it isn't.

"Your television is going to become an interactive delivery system for products and services. You won't even have to buy a new television set. You will be able to buy or lease a box containing a computer/processor and memory chips that will plug into your old television set, allowing it to act more like a computer. New television sets will come with the option of having the computer processing and memory chips built in. You will use a graphic interface, much like the symbols on the screen of an Apple Macintosh or Windows-based PC: a mailbox, a file folder, a pen, a trash can. Choose a symbol with a wireless pointing gadget—kind of a remote control device that functions like a computer mouse—and click a button. Now, you're no longer a passive television watcher, you're an active participant.

"If you clicked the appropriate icon, say a crossed fork and spoon, a list of fast-food restaurants might appear, and you could choose to look at the menus, view a promotional film for each establishment to see if it uses rain forest beef—if you're worried about protecting the environment—and place an order. It will be just like the setup we were discussing for Blockbuster Video. Instead of just receiving and displaying images, digital television will process, store, create, and transmit information."

"I don't see it on the core technologies list," Tanya said.

"None of the cards are. We've got to look beneath the surface to see the core. In this case, advanced computers, distributed computing and digital electronics are driving this technology. Also, fiber optics and advanced satellites are quickly becoming key delivery components."

"What do you mean by 'advanced satellites'?" Glenn asked. "I thought they were all advanced."

"Satellites are like any technology. There's an evolutionary process. Early satellites from the 1960s, even satellites in use five years ago, are pretty primitive compared to what's being put into orbit today.

"In the case of using digital interactive television, many of us will be using direct broadcast satellites (DBS). These are medium-power satellites that require a small, less-expensive receiving dish on Earth. The older satellites you might have used to get your TV signal needed an expensive, large receiving dish because the satellites were low power. The DBS transmission will be digital, giving you CD-quality sound and a super-VHS-quality picture at a fraction of the cost of your old home satellite system.

"Any other questions?"

"Dare I ask about fiber optics?" Lonnie inquired. "You also mentioned it in connection with EDI."

"Yes, but don't get bogged down in these core technologies. I want you to focus on the tool itself. It's the old story about asking for the time of day and getting instructions on how to build a watch. When we're done with the games, I'll give all of you a hefty sheaf of technical information that you can use to broaden your perspective.*

"Briefly, fiber optics will provide a *preferred* ground-based means of interconnecting digital interactive television. Fiber optics form a digital highway carrying all forms of information at the speed of light. When fiber optics is used for communications, you get a four-lane highway. The fiber can simultaneously carry television, telephone, radio, and computer data. If you thought the interstate highway system was important, wait until we finish building this highway."

"Why bother when we've got direct broadcast satellites?" asked Maury.

"DBS is perfect for spanning vast distances, reaching into areas that don't have fiber networks linking businesses with their outlying operations and providing new options for mobile communications," I replied. "Fiber will have its feet on the ground; it's easier to maintain but slower to install than a satellite. Fiber-optic technology will be there handling the more day-to-day aspects of communications. By the year 2000 the majority of homes in America will have a digital path running right to their door, either in the form of fiber optics, direct broadcast satellites, digital microwaves, or a broadband system running on their old home wiring."

"What do you mean by the term *broadband?*" Samantha asked.

"The telephone lines running to our homes were originally designed to carry voice only. Think of telephone lines as a garden hose sending water to your lawn. To send a movie over a telephone line involves transmitting much more data through the same size hose. It would be like trying to put out a three-alarm fire with a garden hose. You would need a wider hose, or, in the case of sending a movie, you would need a broadband width of information versus a narrowband width."

"Got it," Samantha said.

"Okay, Maury, let's move along. Your turn."

*See Appendixes A–D.

"This tool is derived from artificial intelligence, and it allows any-one without programming knowledge to easily and quickly modify or create a computer program. With it, new software packages can be built from prefabricated, pretested 'blocks' of software code, each identified with an appropriate icon or picture. Quickly customizing software to suit the individual user, rather than struggling with a one-size-fits-all program, will be a matter of lining up the icons, with each one representing a different function, like a row of toy soldiers. In the future software can be as individualized as a set of fingerprints.

"Hit me again, Maury."

"This is one that I should take back to the brake and muffler shop that I mentioned earlier when we were discussing pagers. Any noise, whether it is music or a baby crying, generates a sound wave. It can be canceled by generating a mirror-image, out-of-phase wave at the same volume as the original. Truly, the sound of silence."

Brenda Jacobs broke in with a question. "Would that work for sirens? The neighbors who live along the street leading to our emergency room are driven nuts by the noise."

"Eventually it will. But current antinoise applications work better when the sound occurs in a regular pattern and pitch like the roar of a jet engine. An ambulance siren can vary widely according to what kind of racket the driver thinks is needed to get through traffic quickly. Keep following the developments of antinoise technology, though, and when it's ready the hospital's neighbors will be thrilled by the sound of silence. They may build a statue and a small, quiet park in your honor," I said.

I offered the cards to Samantha. "Go for it."

"Neural network computing is often referred to as 'brain inspired.'

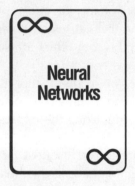

It mimics the workings of the human brain by using an artificial neuron that accepts multiple inputs the way a real neuron does. This feature gives the software the ability to learn through experience and go beyond executing a fixed set of commands. For example, your furnace, using a neural net, could learn over a period of time that you go to bed at midnight every night. Once the neural net has learned your pattern of behavior, you could put it in an automatic mode, and based on its past experience, it would turn down the thermostat at midnight each night. It is also capable of dealing with ambiguity, like converting Glenn's highly ambiguous handwriting into digital data," I said, recalling the doodling he had been doing on a yellow legal pad.

"That I don't believe." Glenn held up the pad for everyone to admire. It was incomprehensible. "The people in the warehouse kid me about it all the time," he said.

"Neural networks may have met their match," I admitted, turning back to Samantha with the cards. "Try your luck again."

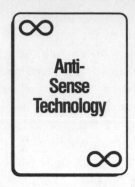

"We're now in the biotechnology area. Antisense makes sense by blocking the characteristics of specific genes. What a breakthrough! By literally turning off a gene that causes an illness or makes a vegetable rot soon after it's picked, this tool offers life-saving and life-enhancing benefits.

"Antisense medicine, or 'code-blocker therapeutics' is a recent important innovation in gene therapy. The power of antisense is that it gives scientists a means of blocking the expression of a specific, faulty gene without interfering with other genes. This can be accomplished either by blocking production of its RNA or by interfering with the action of its protein. Scientists have shown that they can use antisense medicine to block a leukemia-causing virus and to stop replication of human cancer cells without affecting the growth of normal cells. Since antisense blocks the replication of human cancer cells by preventing the expression of genes that are associated with lung and colon cancer, scientists are confident that these two kinds of cancer will become fully treatable using this technique. Antisense medicine is the most promising therapy to thwart replication of the AIDS virus in human white blood cells."

"My industry may have to rewrite all of our actuarial tables on average life expectancy," Samantha said, tapping the face of the antisense card.

"Wouldn't that be nice," I said, reaching for the deck and turning the cards faceup. "Now, I'll choose one."

"Hey that's cheating," Lonnie objected, opening his eyes wide in mock indignation.

"Forgive me. I'm trying to find a card that forms a natural pair with antisense. There it is."

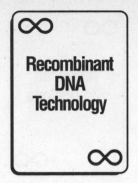

"What's involved here is the mapping, restructuring, and remodeling of the gene code of both plants and animals to eliminate or enhance a specific trait. Just imagine the applications: crops that are insect proof, drought resistant, winter hardy; doctors could predict inherited genetic disorders in their patients and get a head start on treatment; and gene therapy could be used to correct genetic disorders before the symptoms are triggered."

"In my opinion, we could do without that tool, Dan," Tanya said. "First of all, if I have funny genes, I don't think I want to know about it. Secondly, agriculture is in bad shape already. If we start fooling around with plant and animal genes, there is no telling where we are going to end up."

"I understand what you're saying, but remember: *Once the card has been put in the deck, it is going to be played* whether we like it or not. Medicine and agriculture will change the most because of biotechnology. What you don't want to do is bury your head in the sand but instead learn how to use the biotechnological tools to shape your own future in beneficial ways. Now, as for human genes, if I had a genetic predisposition to diabetes, I'd want to know about it so that I could ensure that my diet and life-style were not aggravating the condition and setting my genetic bomb off early. I would also like to know so that I could take advantage of advances in diabetes prevention. And if gene therapy was available to alter my genetic predisposition before I ever developed the disease, I certainly would want to at least consider the option instead of waiting until it was too late."

"Yeah, I see what you mean," she reluctantly admitted.

"See what you come up with again, Tanya," I suggested, offering her the cards.

"Advanced simulations use high-power computers, high-resolution graphics, and digital-quality sound to graphically simulate various situations in two and three dimensions. Are you following me?" I asked.

"I don't know about the others," Glenn said, "but I'm bordering on advanced technological overload; maybe I've spent too much time breathing the fumes of soap powder and rotting avocados."

"In that case, let me back up," I said. "Advanced simulations give us sound and images that possess all of the dimensions and characteristics of the situation that we are trying to simulate, down to sub-microscopic detail if that is what we want to do." Glenn still looked confused, so I tried again. "Remember the television commercial Intel, the PC chip manufacturer, started showing on television back in 1992? It began by showing you a view of a typical desktop computer."

"A PC clone," Glenn added.

"Yes, you couldn't make out a brand," I said. "Then the camera zoomed in on the opening of the disk drive, and you seemed to go inside the computer, whizzing along electronic pathways until you arrived at the central processor, the chip that controls the computer, and you could see the word *Intel*."

Maury quickly added, "I remember they even included a little nonreality to increase drama by having two posts next to the chip that had an electric arc zap between them just over the chip."

"That's right, Maury," I said. "You can simulate the real world or an artificial world, or a combination if you want."

Glenn's eyes lit up as he asked, "Can the viewer ever have control of a simulation?"

"No problem," I said.

"I just had a thought," Glenn said. "You know the old saying 'We can't see the forest because of the trees.' It sounds like I could use advanced simulations to have the viewer see the forest from 30,000 feet in the air and then zoom in to display each magnificent Douglas fir, including chirping birds, the branches swaying in the wind, and the bark glistening with morning dew. If it is an interactive simulation, the viewer could check it out, front and back; even examine the underground root system, if that was preprogrammed into the simulation. And if they'd like to take a look inside the water droplets of the dew—to observe effects of acid rain—they can do that too."

"That's a good analogy," I said. "There are some things that we just can't see, such as shifting wind and temperature patterns around the globe or the molecular construction of an amino acid. The Intel commercial and Glenn's forest are both good examples of two-dimensional (2-D) simulation. This tool, advanced simulations, gives us another set of supersharp eyes—three-dimensional (3-D) simulation called 'virtual reality.' It is an advanced application using a collection of computer hardware and software that allows the user to step into the three-dimensional world, a virtual world, he or she has created and get the visual, aural, and tactile sensations of actually being there. Think of virtual as meaning that whatever you are viewing or experiencing seems to be real, but it's not."

"Reach out and touch . . . a Douglas fir," Glenn said.

Lonnie pounced on that one. "You mean I could design a garden in virtual reality and the client could walk down the pathways and pick the roses—before we ever built the damned thing?"

"It would seem that way to your clients, thanks to data gloves, a special pair of gloves that would transmit the movement of the user's hands to the computer, allowing them to manipulate virtual objects, and a data helmet that is also connected to the computer, allowing them to move their head to see different perspectives in your virtual garden."

"And the next step is to forgo building the garden at all. Just do everything in virtual reality. I could specialize in English gardens in Nome, Alaska," Lonnie said.

I turned to Brenda. "Tell the neighbors to contact Lonnie Weineck when they want to construct that antinoise park near your hospital. . . . Lonnie, I think the *Future you* is beginning to take shape. Pick a card."

"This doesn't seem as sexy as advanced simulations, but it is a powerful tool that allows a computer using a large number of processors to simultaneously attack a problem, reducing the time it takes to retrieve and store data. Think what would happen if we broke a complex problem into seven parts and had each part of the problem worked on at the same time and then fused together the various components of the answer at the end of the process. There would be a multiplier effect in terms of speed and power without having to go to the trouble and expense of building a whole new system from scratch. Parallel processing removes the speed bumps that have traditionally slowed the development of faster, cheaper, more powerful computer technology that can perform the most advanced applications like virtual reality."

I offered Doug the cards. The teacher pondered a moment and chose one that was directly beneath my thumb.

"So what else is new?" he asked, disappointed at what seemed to be a familiar technology.

"Believe me, this is new," I said. "Current cellular phones use analog electronics and in large cities the number of cellular subscribers is close to the maximum limit because of analog limitations. Digital cellular telephones will soon begin replacing analog cellular phones in large cities, allowing for many more subscribers. In addition, digital cellular telephones have no distortion or hiss; they will be able to send computer information without modem and they will have a high level of security. By the end of the decade, a series of Low Earth Orbit (LEO) satellites will be launched into space, allowing anyone on the planet with a digital cellular telephone to communicate to anyplace.

"Try again," I urged Doug.

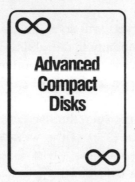

"A good one! Optical data storage is the core technology. Basically, optical memory systems, which are what we are talking about here, use lasers to read information stored in digital form. Examples include optical disks, like music and video CDs (CD-ROM), optical film, bar-code readers and floptical disks." I stopped to explain that floptical disks look like floppy disks but operate more like a recordable CD. "The key benefit of advanced compact disks," I continued, "lies in the technology's ability to store large amounts of information in a relatively small space while at the same time allowing the user to manipulate the data instead of just looking at it, an application known as compact disk interactive. CDI works with current television sets, eliminating the need for separate computers and keyboards." I flashed the card at Brenda. "How about every patient's complete medical history, current hospital records, insurance data, and living will on a three-inch-diameter disk that can be updated at any time?" She lit up, and I could see she was thinking deeply about it.

"Another," I said, as I held out the cards to Doug.

"This is a tool that you will soon be using, Doug. It's a type of software that gives the computer the ability to function as a twenty-first-century version of a ventriloquist's dummy—a high-tech, high-IQ Charlie McCarthy. An expert, or group of experts, working with a software programmer can put together a dialogue-driven program that replicates a consultation between the user and the experts. The experience, the insights, and the skills of the world's finest doctors, lawyers, academic specialists and business advisers are thus available to a far wider audience than ever before.

"In your case, what would you do if you had a problem student and just couldn't figure out how to reach that youngster?"

"I'd talk to my teaching colleagues, maybe go to my principal," Doug replied.

"With an advanced expert system you could sit down and discuss the problem with Robert Coles, the famous Harvard author and child psychologist, any time of the day or night. The software would provide you with recommendations and information drawn from Coles's experience and targeted at your individual situation."

"Deming?" Maury Andrews asked, referring to the famous business guru who pioneered the quality movement. He seemed a little stunned.

"Coles, Deming, Milton Friedman, Edward Teller, Stephen Jay Gould . . . you name them. The expert system technology is there to do the job, and I think many experts will sell their expertise in the form of an expert system. I'm not saying that the software using W. Edwards Deming or Robert Coles is available at this moment. But it is an example of the caliber of expertise that could be utilized."

I gave Maury another chance at the cards.

"Notebook computers will continue to get smaller, thinner, and lighter. The most popular type will be the electronic notepad. Apple Computer calls its electronic notepad a personal digital assistant (PDA). Apple's competitors, in an effort to differentiate themselves from the Apple product, have come up with their own terms. Don't let the variety of terms confuse you; they all fit under the electronic notepad classification.

"Think of an electronic notepad as a highly modular hand-held computer that can be easily customized by the user to fit a particular need. You will choose the configuration when you buy it, choosing either a black-and-white or a color display, pen-based input or a wireless keyboard, or the ability to enter data with your voice. Another popular feature will be a wireless communication system to link up with the telephone system or other computers. The type of wireless communication system will also be an option, such as digital cellular, personal communication network, or a corporate radio link, just to name a few."

Brenda jumped in. "Stop," she demanded. "I thought advanced CDs were incredible. Notebook computers with pens or voice would. . . ."

I gave her the thumbs-up sign in encouragement. "I know. You're starting to see the new big picture. Keep looking."

I asked Lonnie to choose.

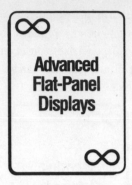

"Several new display technologies will provide us with full-color, flat, lightweight television screens in a variety of sizes. We have all been waiting for this card for a long time, and finally the technology and price are coming together to give us this card. The applications will vary from home television to advertising."

"Operating rooms, too," Brenda inserted.

"Bravo! Operating rooms, too. Optically stored X rays could be called up by a surgeon and displayed on a panel built into the operating room table."

"Grocery shopping carts," Glenn added.

"Hold on! I'll never get through these cards if you guys get carried away now. To play winning hands you need to know all of the options, in this case, the new tool cards. Let's move on to another; we'll come back to creative applications soon.

"Lonnie, do the honors again."

"Desktop videoconferencing is an example of advanced electronic mail. It is similar to a telephone conference call but with live-motion video. Each party uses a personal computer, a small video camera mounted on top of the computer monitor, and a broadband data transmission system such as fiber optics, which allows the user to see and hear the other parties on their computer screen. In addition, each user will have a different-shaped cursor, allowing the user to manipulate the data being displayed on the screen. A new telephone service called Bandwidth-on-Demand will keep user costs low, charging users only for the bandwidth they are using at any one time."

I selected the next card.

"Multimedia computers combine the audiovisual power of television, the publishing power of the printing press, and the interactive power of the computer. Sitting at the screen of his or her PC, a user could decide to learn about Dr. Martin Luther King. A menu pops up, and the user selects the text of a biographical essay from the choices. At the reference to King's famous 'I have a dream' speech, the user can elect to watch a news clip of the speech and listen as the civil rights leader speaks on the steps of the Lincoln Memorial. Afterward, he or she can browse through still photos of King's life and listen to the comments of Andrew Young, one of the civil rights leader's confidants.

"Because of rapid changes in the marketplace, businesses of all sizes will need to focus on the retraining of their work force. A well-trained work force is not as valuable as a work force that is capable of being retrained again and again. To keep costs down, businesses, governments, all organizations for that matter, will use multimedia as a vehicle for just-in-time training."

"I think I know what you mean by 'just-in-time training,' Dan, but could you define it for me?" Glenn asked.

"I'm glad you asked," I responded. "Instead of taking employees off the job for training every year or two and teaching them what you think they will need to know, you can let the employee tap into the interactive 'virtual classroom' anytime they need new knowledge. There are a variety of ways to deliver knowledge to remote sites, including satellites."

I threw out another card.

"Telecomputers are a combination of a Touch-Tone telephone, answering machine, fax, and home computer with a flat-panel display using an electronic pen or touch screen for making selections or inputting information. They will provide services such as electronic newspapers, information-on-demand, faxback-on-demand, and electronic shopping; high-end models will offer two-way picture-phone functions. You will have the option of receiving your interactive television signals through your telecomputer. Multimedia voice-mailboxes will be another popular option."

"What do you mean by faxback-on-demand?" Tanya asked.

"Using a traditional Touch-Tone phone or a telecomputer, you can dial an information provider, such as your local newspaper, and the electronic, human-sounding voice will ask you to press one if you want the latest weather map, two for the latest sports statistics, and so on. As soon as you press the number you want, you will dial in your fax number, and the fax will immediately be sent to that number. Since your telecomputer is also a fax and it also has a screen, you will be able to see that a fax has been sent. You will touch the fax symbol with your finger to see the fax on the screen, then decide if you want to print it out."

"I sure like the idea of seeing the fax before printing it out; I could save a lot of paper by not printing so much," Tanya said.

"Brenda, give it another go," I said, offering the cards.

"Digital imaging will help us dig out from under the mountain of paper that we're creating. It will take words, pictures, charts, and graphics and turn them into a digital form that computers can read and manipulate. Hand-held and desktop scanners are going to be widely used. Those clippings from the morning newspaper won't just sit there and turn yellow. A few minutes with the scanner and the information is in the computer.

"Automatic, large-capacity, self-sorting scanners will give offices and large organizations the same capability on a high-volume basis. And electronic digital filmless cameras will capture, store, manipulate, and display photographic-quality images, from snapshots and passport photos to medical X rays.

"Back to you, Maury. I have a hunch the odds favor the selection of a card you and I talked about the first time we met."

Personal Communication Networks

"No, wrong one. I'll give you another try, but let's talk about this tool first. PCNs are going to put a major dent in the use of the traditional pay telephone. The actual device you will use will be called a personal communicator, and, as with the electronic notepads, competing companies will tend to use their own term to describe their version. Regardless of the brand name, all personal communicators will have several things in common. They will be less expensive and usually a little smaller than the smallest cellular phones. The lower price is the key advantage over a cellular phone. They are not meant to replace cellular technology. Instead they are meant to be an inexpensive alternative for wireless communications. Think of them as a cross between a pager and a cellular phone. To keep costs down, some units will not be able to receive a phone call as on a cellular phone. Instead, the person wanting to contact you will send information like you would on a pager. Using its little screen for reading the message, you can then place an outgoing call as you would with a cellular phone. The outgoing call will go through a ground-based network of small radio receivers that will be linked to all worldwide telephone and data networks. It will send and receive digital information, thus providing interference-free communications, and it will be far less costly per call than the existing cellular service.

"Now you get a second try for that card I thought you might draw the last time, Maury."

"That's it. There aren't many cards left. I asked you how your company was using fuzzy logic technology. Remember? Well, this tool is going to be very useful to you and other manufacturers. Fuzzy logic is a form of math that lets computers deal with shades of gray. All computers work on a 'yes-no' principle. Fuzzy logic adds 'maybe.' This allows computers to operate at a higher level of abstraction and handle conflicts or contradictory commands.

"With a yes-no principle, a machine is either on or off. A light bulb used to be shining brightly or it was turned off. The innovation of the electrical rheostat gave us the option to adjust the bulb's luminosity. Fuzzy logic does the same thing. Instead of an elevator or subway car jerking to a stop, it glides smoothly until it is in the proper position.

"Four more cards will complete the new deck," I said, selecting a card.

"The majority of today's small and medium-size manufacturing plants, as well as many large operations, have not integrated the 'internal' customers' computing and control systems. The various functions tend to exist as independent islands of incompatible information, with little communication and poor coordination."

"Hold on, Dan," Maury interjected. "I know what you're talking about when you use the term *internal customers,* but I wonder if the others do. The simple way to understand an internal customer is to figure that when my engineers produce a new product design, their internal customers are our manufacturing and production people." Maury looked at Doug to see if he understood. The teacher nodded, and Maury continued: "If the engineers design something that production can't build or thinks is a piece of garbage, I've got a problem. Once the corporation's various internal customers are out of synchronization, satisfying the external customer is Mission Impossible."

I pointed at Maury to indicate that I agreed completely. "But of course that never happens," I said facetiously.

"No more than a couple of times a year," Maury replied and shook his head.

"CIM to the rescue," I said. "Computer-integrated manufacturing is a form of distributed computing that allows the total integration of a manufacturing enterprise by bringing the internal customers into alignment, from engineering and planning systems at one end to the devices controlling the physical operation on the assembly line at the other. By combining all computer hardware and software, regardless of brand, with smart, computerized machines, manufacturers can cut time and costs by eliminating redundant procedures.

"Then there were three," I said.

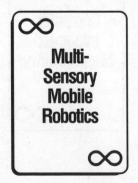

∞

Multi-
Sensory
Mobile
Robotics

∞

"Let's start this one with a simple definition of a robot. Basically, a robot is a computer that can move. Robots are different from other forms of automation in that they can be reprogrammed to do a completely different job. The assembly robots in use during the eighties and early nineties never lived up to their expectations because they had few sensors and were only mildly adaptable. The new generation of advanced robots possess multiple sensors and increased mobility. Industrial robots equipped with multiple sensors and the ability to move to different locations within the plant will provide the missing key ingredient manufacturers will need for success in the future—flexibility. With the ability to design infinite variation, they will perform a variety of tasks from providing security to mowing lawns. We haven't seen anything yet when it comes to robotics.

"And the final pair."

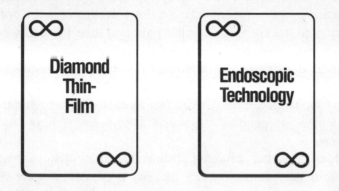

"Diamond thin-film: By depositing an ultrathin layer of diamond film, only several molecules thick, on a surface such as a blade of a knife, a razor, or even a loudspeaker, you create a surface that has the properties of a natural diamond. The knife or razor will never get dull, and the stereo speaker will yield higher highs and lower lows without distortion. Diamonds are the hardest substance known and can, therefore, deal with power and heat levels far beyond those that would destroy the electronic properties of most other materials. Diamond thin-film will provide new applications for computers and sensors working in very hot or cold environments.

"Endoscopic technology: The primary, and most dramatic application, will be surgery. In endoscopic surgery, a long, hollow tube (endoscope) is inserted through an incision less than an inch long.

Fiber optics inside the endoscope carry a bright light from outside the patient down to the tip of the endoscope. A tiny television camera at the viewing end of the endoscope transmits pictures to a television screen, which the surgeon watches while performing the surgery with a small surgical tool. Using this less invasive surgical procedure, patients recover within days rather than weeks, saving medical expenses. In addition, the tiny incisions that were made do not have to be suture-closed and often are simply covered with Band-Aids. There is less pain, swelling, and scarring. By the end of the decade, 75 per-cent of all surgery will be endoscopic.

"There are other applications for the endoscope, including replac-ing periscopes in submarines and allowing mechanics to inspect the innards of working jet and internal combustion engines."*

"Since I'm in a hospital all day, I tend to keep up on medical advances we are involved in. Using laparoscopy on someone getting a hysterectomy, our surgeons can cut the average six weeks of painful recovery time to one week of mild pain and then back to work. How is this different from the endoscope?" Brenda asked.

"Good question," I responded. "Endoscopic technology can be adapted for a wide variety of operations. Think of laparoscopic surgery—*laparo* is a term derived from the Greek, denoting the 'loins' or 'abdomen' in general—as a specialized variation of endoscopic technology."

When I finished, I moved around the table, giving a pack of the twenty-four new cards to each player. "Between the new rules and these new tools you are now playing with a full deck. The future is yours to shape, but next you're going to have to learn how to play with those cards. There's no way around it."

*See Appendix C.

4

Opportunity and Obsolescence

Shuffling the Decks

Maury Andrews had been doing his homework since our last game one week earlier. He arrived at my house with four tool cards preselected from his new deck.

"I kept turning over the new cards until something clicked," he told the group. "My secretary must have thought that I had gone round the bend. There I was, sitting at the desk all alone, arranging the cards in a row running left to right. I'd turn one over, stare at it, shake my head, and try the next."

"I hope she doesn't rat on you at the next board meeting," Lonnie said.

Maury frowned. It must have been a disturbing thought, but he continued with his story. "I probably went through the deck four or five times before I turned up this card again."

Maury picked up the card from the table where he had tossed it out for us to see. "This card was saying something to me, but I didn't know what it was right off."

He paused and tapped the card on his forehead. "At first I thought that the card didn't belong in my deck. It seemed too specialized for my mainframe manufacturing operation, something a third party would produce to go with my systems. If it's going to allow my customers' employees to custom-design sophisticated computer programs by moving pictures representing different functions—"

"Don't they call those icons, Maury," Brenda suggested, interrupting Maury's statement.

Maury smiled, "They sure do, and if the sophisticated programs that run on my mainframes could be easily customized by any individual user, that might cut into our profitable software support services. Object-oriented programming didn't seem to be in our best interest."

The group was fascinated by the revealing statement. We sat there in silence and waited for Maury to continue.

"I was about to flip it into the wastebasket when it occurred to me to pull one of the cards from Dan's new rules deck."

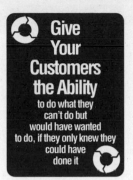

Give Your Customers the Ability to do what they can't do but would have wanted to do, if they only knew they could have done it

"That's when it hit me. The two cards together made the right combination." Maury slapped both cards on the table side-by-side. "Object-oriented programming can make the mainframe as easy to customize and use as a PC. Those monsters that I make can take a lot of time for my customers to learn and then customize for their applications."

"You're onto something, Maury," Brenda Jacobs interjected. "At the hospital we've cut way back on our plans for using our old mainframe system because we thought it couldn't match the ease of use that networked PCs can achieve. Using the mainframe is so demanding for a large number of key hospital personnel. They say they'd rather use smoke signals or semaphore."

The CEO appreciated the encouragement, but I don't think he liked hearing his mainframes judged so harshly. "Anyway, Brenda, if I said to you, 'Go ahead, use the mainframe like a Macintosh, and any of your people could quickly tailor an application to suit their individual needs,' I'd be giving you the ability to do something you wanted to do but didn't know you could do," Maury said, leaning back in his chair. "The hospital wouldn't end up junking the mainframe, and we could go ahead on developing the next-generation operating system and new hardware to go with it."

Doug Stedman spoke up. "Isn't that just a temporary fix, Maury? Your competition will play the object-oriented programming card too, and you will lose the edge."

I thought the comment might throw Maury, but he actually looked rather pleased. He reached for the deck of new rules and pulled out another card.

"We will design our preprogrammed objects, or as Brenda pointed out, icons, to work with our current mainframe system and with PCs, allowing us to tap into another huge market. Change is built into the plan," Maury said.

I had to jump in to congratulate him. "A winner! The card game is meant to stimulate your creative applications of new technology. The true power of the cards comes from mixing and matching. If you had just used the object-oriented programming card to give your cash cow another few years of life, it wouldn't be nearly as effective. . . ." I stopped and pulled a card out of my shirt pocket.

"It should be up my sleeve," I joked. "This one is a new rule that I haven't introduced yet. Take a look."

Maury hesitated. "I wasn't trying to make the mainframes obsolete; I was out to keep them going," he said.

"You said something about the new-generation operating system and hardware, didn't you?" I asked. "What do you have in mind?" In reply, Maury put his second new tool card on the table.

"Dan and I talked a little about parallel processing when we first met," Maury said. "Parallel processing computers will allow my customers to analyze large amounts of data at much higher speeds than high-end mainframes and at a lower cost. Other companies have already started to play this card; it's about time I start playing with a full deck. We could continue to offer our customers mainframe capabilities without manufacturing a traditional mainframe. We could offer a large-capacity parallel system for our biggest customers, and a system for networking PCs in parallel allowing our smaller customers to tap into the power of parallel processing when they need to."

I nodded my head. "Look at the new rule card. Aren't you rendering your cash cow obsolete?"

"Yes, I suppose I am, come to think of it. But I'm continuing to milk it while I get my parallel processing act together."

"Wow! If you didn't do that," Glenn nearly shouted, "the networked PC guys using object-oriented programming *and* parallel processing would eat your lunch."

"Good-bye cash cow; moo-ve to PCs," Lonnie wisecracked.

The Old and The New

As the laughter subsided, I moved to the head of the table to make an important point. "The old cards don't immediately and automatically lose all their value when the new cards are put into play," I said. "There is a transition period during which the old tools and rules still have effect but their potency is gradually diminishing." I gestured toward Maury. "Mainframe technology is already in the later stages of transition. Your current customers are still using your products for a variety of reasons—they need the muscle that networked PCs don't offer as yet, can't afford to make the switch, and they have a large financial investment in existing hardware. In addition, they know they will not have to retrain people if they don't change anything. Therefore, the old mainframe card will remain in play for quite a while. The market, however, is mature, and both you and your customers need new blood. Remember, status quo is now defined as rapid change!

"The first mainframe system was introduced in 1964, the year you got your MBA from Harvard, Maury. Remember the new rule: *If it works, it's obsolete.* And those mainframes have been working since 1964."

"So, Dan, what Maury has to do is phase in the switch to parallel processing?" Samantha asked.

"Yes, to start playing with new cards while the old cards are still in use. Object-oriented programming allows him to stick by his old customers and products while the transition is taking place. I assume that the object-oriented programming that Maury develops will be proprietary and, as he said, fully compatible with the new operating system. Meanwhile, new customers will be lining up eagerly to buy into both parallel processing and objected-oriented programming backed up by the company's solid reputation and long tradition of service. It's a product they can't get anywhere else."

"I seem to be playing the skeptic here tonight," Doug said, raising his hand for attention. "If Maury goes heavily into object-oriented programming and parallel processing, won't the little guys do it too and then beat him on price and options and service?" Doug looked around the table and shrugged apologetically.

Tanya was the first to reply. "But Maury now has a head start. By the time the competition catches up, and if he keeps rendering his cash cows obsolete, Maury can stay in the lead forever."

"Providing he uses this card," Brenda said, rummaging through the deck of new rules. She quickly found what she was looking for.

"You've got it," I said. "Not change today and status quo tomorrow. Change today, tomorrow, and forever."

"That's the trouble," Maury said. "I wonder if I'm going to be able to keep coming up with new cash cows. For all I know, parallel processing and object-oriented programming are the end of the line."

I shook my head. "No way. Object-oriented programming has been around for more than twenty-five years, and it is just now coming of age. What do you suppose is being dropped into the mouth of the technological pipeline today that's going to be tumbling out the other end in a quarter of a century?" Maury and the others looked blank. "If you start playing with today's new cards, mixing and matching, picking up on whatever new cards the future deals out will become second nature. You'll be able to spot them coming years before anyone else.

"By having a basic knowledge of the twenty core technologies that I introduced last time, you are acquainted with the ingredients—the flint and steel—that will be sparking technological revolutions well into the next century. It's like the old movie marquee slogan: 'Watch This Space.' Watch those core technologies, and the cards will start popping up. For example, voice recognition software, which comes from the core technology artificial intelligence, is close to entering the tool stage. There are twenty-four of them today. In ten years there could be fifty."

Price and Time

"The mistake is to take a wait-and-see attitude. If you wait until the technology is cheap and abundant, it's already too late," I said. "One indicator of a technology's maturity is price. At first, as it is being developed, the price is usually extremely high. Once R&D costs are covered, production increases and competition sets in; then the price starts dropping. When an alternative technology appears the price often plummets."

"That happened to the 286 and 386 computer chips," Samantha observed.

"Because the 486 and RISC chips were right behind and are closely pursued by the next generation," I said.

"So it is the reverse of the stock market, where you want to buy low and sell high. Go after the technology with the high price tag and dump out when it falls off," Maury Andrews said.

"Price can indicate two things, Maury. One, that the technology is in the earliest stages of the life cycle. Or, two, that it is approaching obsolescence. As the demand falls off for the technology, maintenance, spare parts, and trained personnel become costly."

"Sounds like you're describing my 1962 Studebaker Hawk," Glenn said.

"And that Hawk would probably command a pretty healthy price these days," I replied. "Part of the price for the vintage car is collector's value. Another part is the previous owner's attempt to recover his investment in upkeep. Vintage technology carries a high price tag because its manufacturer is losing money keeping the antiquated assembly line open—sound familiar Maury?—suppliers are charging a premium for components they'd rather not handle because of the limited market, and customer complaints and service demands are on the rise since that end-user is driving the technology beyond its capabilities in an attempt to keep up with the more technologically advanced competition.

"The other thing about price is that it measures time. Early in the life cycle, the high price indicates that the technology is in its infancy. While that may mean growing pains, it also suggests that it will be around for a while. As the price declines, so does the remaining life expectancy."

Doug was nodding, and I paused to let him comment. "You pay a bargain price, but you could be getting technology that is approaching obsolescence, and you don't even know it," he said. "I remember that as CD players started to take off, the price for LP turntables and cartridges really fell. I bought a really nice turntable and a year later couldn't resist getting a CD player. Now I hardly ever use the record player. Some bargain!"

"What you're saying is ominous," Glenn said. "I've got a huge collection of LPs. When my turntable or cartridge wears out in ten years, I'll have to pay through the nose for replacement equipment. They'll be specialty items."

The exchange was just what I was hoping would develop within the group. I wanted them to reduce technological complexity down to familiar products and situations. "Now you're seeing how the price and time lines intersect," I said.

"I can confirm that," Maury said. "We've had to cut our prices to keep market share. I thought it was a measure of the level of increased competition. You're saying, Dan, that it reflects the obsolescence of my technology."

"Precisely. Your competition is in the process of rendering your cash cow obsolete. Every time the price line drops, the life expectancy diminishes. The cash cow is taking another step toward the slaughterhouse."

"Dan, you keep using the terms life expectancy and life cycles. We're talking about inanimate objects, though. Does technology really go from conception to gestation, birth to infancy, adolescence to middle age, old age and death?" Tanya Conte asked.

I went to the flip chart and started writing. "Let's take a look at the five key stages a new technology needs to go through in order to make it to market, starting with discovery."

Technology Development Time Line

Stage 1	**Discovery** Development of a new idea or theory
Stage 2	**Observation** Possible practical applications are identified
Stage 3	**Feasibility** Proof of technical practicality
Stage 4	**Development** Market-oriented experiments
Stage 5	**Production** Introducing the new process or product in the marketplace

I said, "Every technological development travels down that time line. Let me explain how it works. One fine day the inventor slaps his forehead and shouts, 'Eureka! I've got it!' That's the discovery date. The period of observation follows while the inventor and his backers and his rivals take a look at the discovery with an eye to identifying possible practical applications. The third stage is feasibility, when it's determined that the technology will actually deliver those practical applications. Next comes development, a period of trial and error, or what the military weapons procurement people call 'Fly before you buy,' to get the technology ready for the market. Finally, there is production, and the technology heads off to the customer."

I turned the page. "Let's take a look at a chart that will help you to see the relationship between several key factors in the life cycle of a new technology once it reaches the production stage."

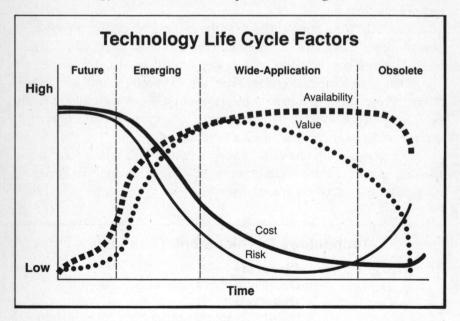

I continued, "Let's define future technology as a technology that has just become available, but there is no current, proven application. You might say, it's looking good, I'll keep my eyes on that one.

"Emerging technology will include technology that has found applications in other industries but has no current applications for your specific industry yet.

"Wide-application technologies have found widespread acceptance in many areas, including your industry.

"Obsolete technologies have failed to adapt to rapid changes in the marketplace, or they may have been rendered obsolete by a more cost-effective alternative.

"Time is the most important factor in determining the true value a technology has. Most of us tend to use *cost* as the most important factor in determining the value of a new technology. This is a great mistake. We must look beyond the cost to see the true value.

"Take a look at the *value* line. This should be the most important factor to consider. The true value of a new technology to you or your organization, regardless of size, lies in the competitive advantage you can leverage from it. This will be the highest when few in your industry know that it exists. Most will apply new technology to do what they did before. Later, in the wide-application time frame, is when you typically will see creative applications emerge. Once your competition begins to creatively apply the technology, its true competitive value will begin to drop.

"What happens most of the time is that new technology is virtually ignored until it reaches the production stage. In the past that was no problem. Why? Take a look at this chart."

TECHNOLOGY	INVENTION	PRODUCTION	DEVELOPMENT TIME
Fluorescent lighting	1852	1934	82 years
Radar	1887	1933	46 years
Ballpoint pen	1888	1938	50 years
Zipper	1891	1923	32 years
Diesel locomotive	1895	1934	39 years
Cellophane	1900	1926	26 years
Power steering	1900	1930	30 years
Rockets	1903	1935	32 years
Helicopter	1904	1936	32 years
Television	1907	1936	29 years
Kodachrome	1910	1935	25 years
Transistor	1940	1950	10 years

I continued, "The first entry really tells a tale. Eighty-two years elapsed between the invention of fluorescent lighting and its production. Talk about cursing the darkness. The inventor was dead before his brainchild came of age.

"Take the invention of the zipper as an another example. Thirty-two years of development time. More than enough for the garment industry to effortlessly make the switch away from buttons and for luggage makers to give up on buckles and straps. Change came slowly. But it came, and fortunes were made by individuals who were far-sighted enough to see what was ahead by identifying core technologies that had been invented decades before.

"Today, however, the period between invention and production is getting shorter and shorter. Look at the last entry on the chart, the transistor. The interval was down to ten years. It took roughly fifty years to develop the telephone and only four to travel from the first controlled nuclear chain reaction to dropping 'Little Boy,' the first A-bomb.

"In the past, individuals and companies actually hid in the shadows between the oncoming and receding technologies. Middle-aged workers could count on retiring before they had to make major adjustments. Meanwhile, the firm's ownership or management would be regularly hired and fired and put out to pasture, bringing on incremental shifts in policies and practices.

"When we first started playing with the new cards, I said change has changed. It's much faster and nobody can hide in the shadows."

"Dan, you also told us that the future has already been invented but that we just can't buy it off the shelf yet. Are you saying now that the speed of change has eliminated the possibility of even getting a little bit of an edge by tracking and anticipating emerging technology?" Lonnie asked.

"Not completely. For one thing, it means you have to watch the emerging technology particularly closely because you're not going to have ten or twenty years to adjust. The time lag will be more like two or three or five years. Remember the rule I just applied to Maury's 1964 mainframes: *If it works, it's obsolete.* As soon as technology reaches the wide-application stage, it must be considered to be outmoded. We can actually redraw that last chart in the form of a time line this way:

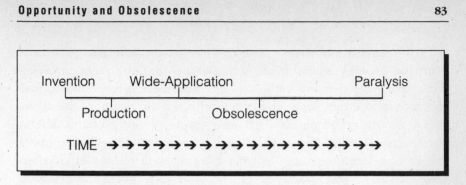

"The mistaken assumption is that we have years, if not decades, between production and paralysis. Actually, it may only be a few years. As a result, we end up managing creeping obsolescence and creeping paralysis. The only option is continual reinvention."

The group sat in silence, staring at the time line until Doug spoke up. "I'm looking at that chart and substituting people for technology—my students. As soon as they graduate from high school, leave the production stage, and go out into wide application of their knowledge, they are obsolete and headed for paralysis." He paused and shook his head. "That's a pretty bleak outlook."

"It is, Doug," I said, "as long as we persist in thinking that our students *graduate*. The word implies an ending rather than a beginning. I prefer to use *commencement*. Re-invention, when applied to people, might be described as re-becoming, which is a lifelong process. None of us is ever going to graduate."

"But how do you teach—what do you teach—if things are in such a state of flux?" Glenn asked.

"The teaching curriculum has to prepare students for both the present and future reality. Not the reality of 1990 or 1980. It has to be constantly changing and evolving. In that way, and only in that way, will students develop a mind-set that welcomes change and readily adapts to it."

Fathers and Sons

Maury cleared his throat. "My son, Greg, is a good case in point. He went through high school and college without ever settling down and focusing on a career. At his age I knew what I wanted to do." He looked at Doug and concluded, "The schools are failing our children."

"Whoa, hold on Maury," I said. "That may be true but not necessarily for the reasons you think. When you were Greg's age, knowing what you wanted to do with your life was the expected norm, and let's face it, you had fewer choices. It reflected the reality of the 1950s and '60s. You could plan a thirty-year career. When a young person says to me, 'I don't know what I'm going to be when I grow up,' I say, 'It's no wonder to me; half the jobs that will be available in ten years haven't been invented yet.' In your day, people specialized early. In this new, rapidly changing world, Greg's better off with a good broad foundation of experiences and education to build on. He is not much different than the millions of other young men and women who are driving their parents crazy by seeming to drift from job to job, lacking commitment. He's probably using your values to judge himself, and he feels like a failure. Those values were great for a time that no longer exists."

Samantha pounced on that immediately. "How old is your son, Maury?" she asked.

"Twenty-five."

"We're the same age. Does that mean, Dan, that I've made a mistake to get into insurance and plan to make it my career?"

"Let me answer the question with a question: Why did you choose insurance?"

"I had a high school guidance counselor—a woman—who said that insurance offered a lot of opportunities for women. She said the industry tended to be dominated by men despite the fact that more and more women are in the work force, raising children on their own, leading independent lives. Traditionally, men sold life insurance policies to other men. Now and in the future, women are going to be making those decisions for themselves."

"Good for your guidance counselor and good for you. You are making change your best friend. You're not going to have any trouble adjusting to the future. Greg Andrews, on the other hand, wasn't as lucky as you were. He didn't have that future-oriented, change-oriented adviser. And to that extent he was failed by the schools. Probably the future looks scary and not at all promising. Therefore, he doesn't want to make commitments and get tied down. By staying loose, though, if Greg can be persuaded that the future is bright, he is in a better position to seize his opportunities than young people who blindly get into a career track just because they think that is what they're supposed to do based on their parent's experience and expectations."

"But Greg has had four different jobs in the last three years," Maury objected.

"And twenty years ago that would have been a problem. He would have been tagged as a floater, as unreliable. Today, there's no stigma within his peer group, and there shouldn't be with you either. He is adding depth and breadth to his resume," I said. "He's going to need wide experience and a variety of skills to succeed in the future. Would that have happened if he had spent the last three years in a highly specialized job in your company?"

Maury pondered the question for a moment. "I wouldn't want him there," he said. "Too much turmoil and uncertainty."

"Exactly. How about spending the next thirty years in your company?" I asked.

"I don't think anybody will ever again spend thirty years cradle to grave in our operation."

"That's true for the entire economy. Job security has given way to job adaptability. Greg, consciously or unconsciously, is cultivating his job adaptability."

I think Maury was beginning to see what I was driving at, but he still needed to be convinced. "He doesn't have sufficient experience in any one area to provide a foundation," he insisted.

"That's the reason I'd hire him. Greg hasn't invested in the status quo," I said. "He is seeing the job from a totally new perspective. He is the ideal change agent. The young and the old make for perfect partners. We've got to end the war between the generations. Combine a young person's enthusiasm and lack of preconceived ideas and old habits with the valuable skills of an older, more seasoned person who's openminded enough to listen, and you've got a winning team."

"We do that at the hospital," Brenda Jacobs said. "It's a mentoring or buddy system with pairs of veteran and neophyte nurses. Sometimes I think it's the most beneficial to the one who is the mentor. They get a charge out of working with the kids."

"I had a similar experience as a young science teacher," I said. "My first year of teaching I was teamed up with a teacher who was close to retirement after having been at the school for nearly twenty years. She helped me to quickly learn my way around. In turn, she

was open-minded enough to try what she considered to be rather radi-
cal ideas, and together we were able to do some exciting things in the
classroom."

Chain Gang

"All right, Maury, we've seen two of your new tool cards. How about
the third," I suggested.

"Let me guess before he shows us," Glenn requested. "I know the
one I'd play." He searched through the new deck and laid out a card.

"You've got it," Maury agreed. "The more I think about EDI and
its ability to link up all the component parts of the manufacturing and
sales process—suppliers, manufacturers, distributors, customers—the
more sense it makes. Our salespeople are forever screaming about
delays in shipping equipment, and most of it's caused by our suppliers
being slow to get us the parts to build an order. Before playing the
card games, I thought EDI would work well only in a large retail store
like a K Mart, or Wal-Mart. Now I can see how we could adapt the
concept to work in both our PC and mainframe divisions. By alerting
our suppliers to all our PC sales as they happen and by electronically
notifying them of a pending mainframe sale as the contract is being
prepared, they could begin to prepare to ship the materials and
arrange for the overtime that would be necessary instead of waiting
until the last minute." Maury reached for another of his rule cards.

Enter the Communication Age

"EDI really makes that possible," he said. "It's going to take a lot of the guesswork and blind flying out of the process."

"Electronic data interchange is going to destroy the old 'one hand doesn't know what the other is doing' bugaboo. It's going to forge new relationships in the business chain," I pointed out. "In the past we only communicated internally. In the future, we'll reach out and extend the communications network to include the entire world of business. I told you that The Gap stores are using EDI, and so is Wal-Mart. I had a client, the CEO of a large cookie company, call me to ask about EDI. He said that he was in danger of losing his contracts with Wal-Mart because they were insisting that he join their EDI network or get lost. The executive wanted me to explain what electronic data interchange was all about. It was a mystery to him. I started off by telling him that with EDI, every time a customer in Abilene buys a dozen of his chocolate chip cookies, the information will show up on one of his computer terminals as well as on the terminals of the operation that supplies him with those little chocolate bits. The knowledge of day-to-day transactions in all retail outlets would give him the ability to cut time, inefficiency, and costs. Concepts like low inventory control, just-in-time delivery, and lean, flexible food manufacturing cannot be accomplished without the right knowledge, and EDI provides the needed information. Fifteen minutes later the guy was kicking himself for not having EDI up and running that very day."

I could see that the possibilities of EDI really appealed to Glenn. "Let's face it," he said, "we're all somebody's customer. My company distributes tons of brownies every year. And I probably handle that CEO's cookies, too. Why not tie all the customers together, from the farmers and fertilizer makers, the granaries, millers, and bakers, the plastic wrap and oven manufacturers, to the trucking companies and the accountants?" Glenn shook his head in wonder. "Dan, you mentioned Ben Franklin earlier; when they signed the Declaration of Independence, Franklin said, 'We must all hang together, or assuredly we shall all hang separately.' It seems to me that without EDI we're all hanging separately. Literally, hanging."

Maury jumped back in with just as much enthusiasm. "I'm just worried that my competition is going to get this setup before I do."

"Anybody who doesn't have EDI is going to get murdered," Glenn shot back.

"Absolutely. I could probably set up a variation of an EDI system to inform me when the end-user of my computer systems has increased his business and is beginning to need more hardware." Maury paused to reflect, but noticed that the others were watching him closely. He rifled through the cards and slapped down another.

"My computer systems could use a neural network to learn the patterns of wear and tear my customers put their systems through. We could tie this information into a type of customer-use EDI system, using the information to improve our next generation of hardware and to alert the user to parts failures before they happen. This would save them downtime, and in most cases that's more expensive than the repair," Maury said.

"How about approaching the customer about buying a whole new system with more capacity to handle the increased sales volume they will probably experience," Lonnie suggested.

Maury made a thumbs-up sign. "Of course, it's a natural. The customer isn't caught off guard and unable to handle his new business opportunities, and we are in the position to book new orders to fill the demand. Talk about a win-win situation."

The excitement was contagious. Brenda had an idea about networking the pharmaceutical companies, hospitals, labs, and surgical supply houses together. Samantha jumped in with an EDI application that would link hospitals and insurance companies. Tanya wanted to be part of Glenn's brownie industry EDI to supply butter and milk. Doug saw EDI as a way to solve the perennial problem of book and supply shortages at his school. "Using EDI, I could link with colleges and employers to track my students after they leave school," he added. "We could see how they perform and adjust our courses accordingly."

"You're all using a new rule card without realizing it," I said. "Anybody want to guess which one it is?"

Maury reached for the deck. "I know because I'm going to combine it with my next tool card," he said a tad smugly.

"Not bad! EDI is a better path to the customer. In fact, it is a superhighway to the customer, giving them what they want when they want it. And once they start traveling in that particular fast lane, there is no going back to a non-EDI environment."

Maury barely let me finish before he tossed out his fourth tool card.

"You challenged me, Dan, about not having frequent contact with my customers. This card lets me do it without requiring extensive travel or interminable meetings," Maury said. "It is a much better path for me and the customer. Plus, I see another advantage. I'm going to plug my customer service people into desktop videoconferencing because it is one area a personal, face-to-face contact can really make a difference. If a customer has a problem, he wants it to be handled by another person, not a disembodied voice on the other end of a tele- phone line."

"You could also use desktop videoconferencing internally to sup- plement face-to-face meetings and to allow you to meet some of your employees who otherwise might never get a chance to actually look the boss in the eye," Doug suggested. "If I could do that with my school superintendent, we might actually begin to start ironing out some of the problems at school."

"Dan, does it just display the video images of the people I'm com- municating with, or can data be displayed as well?" Samantha asked.

"You can split the screen into windows and display documents such as service contracts or pages from a repair manual. Each party would have control of a different-shaped cursor so that each could underline or make changes in the documents."

"That means I could put an insurance policy up on the screen and have a videoconference about it, analyzing the document line by line?"

"You've got it," I said.

"Boy oh boy, that would allow me to double my sales calls each week by eliminating travel time," Samantha exclaimed.

"You and Maury are using this new rule card," I said as I tossed it out.

"Time is the currency of the nineties and the twenty-first century. We're going to have to learn how to actually make time with technology, and that's what Maury is doing by using desktop videoconferencing to meet with his customers. Samantha's doing the same thing," I explained.

I turned toward Tanya. "Tanya, didn't you say your older brother went into the advertising business instead of staying on the farm?"

"Yes, I did, and he would immediately see the benefit of this tool. The agency could have what I think they call 'creative meetings' with various writers, artists, and idea people using desktop videoconferencing. The copy could be changed as each participant got a brainstorm; different kinds of art could be tried out. He'd love it."

"Another enabling technology, digital video interactive (DVI), can be used to compress video so that it can be stored in your hard drive or sent down a phone line, and those ad folks could produce their TV spots with music and action video during a desktop videoconference."

Double Shuffle

It was getting late, but I wanted to give the group one more illustration of the power that can be achieved by literally shuffling the tool and rule cards to produce combinations that would not otherwise suggest themselves.

I gathered the cards together in a single pile and asked Doug to cut the deck. I shuffled them and Doug cut them a second time.

"Here are two cards from the deck that Maury put together," I said. "They happen to be his first and last cards."

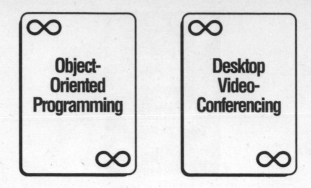

I asked, "Can we make those two cards work together? Can we turn them into two of a kind and win the poker hand?"

At first, there was no reaction.

"Oh!" It was Lonnie. "Use desktop videoconferencing to train your customers, Maury, to learn how to use your new object-oriented programming software." He leaned back in the chair and smiled. "And since you've got EDI, run your desktop videoconferencing down the same fiber-optic network."

"Or over direct broadcast satellites," I suggested, feeling a little like the teacher who was being eclipsed by his students. "For that matter, parallel processing, another of Maury's tool cards, is going to be used to give EDI the capacity to analyze vast amounts of information at blazing speeds. There's another card combination."

Any hope of an early evening went out the window. For the next hour and a half the players were raising and calling by trying out various combinations of cards. They would have been there all night if I hadn't turned off the lights.

5

Picture Power

Keep Your Eyes on the Cards

I was still setting up for the Wednesday-evening game when Glenn Clayton rang the front doorbell.

"Sorry I'm so early," he said. "I wanted to talk a little before the others arrived."

I told him there was no need to apologize and suggested we go sit in the living room. Glenn settled into the sofa with a barely audible sigh. He looked like it had been a tough day.

"What's up?" I asked.

"You know how excited I was the other night about EDI. The next day I had lunch with my father, who, as you know, still technically owns the business. I told him all about EDI and how we could really move out in front of the competition if we went in that direction. It was like talking to a brick wall. He just shot me down."

"You've got the flip side of Maury Andrews's problem," I observed. "Another casualty of the war between the generations."

"The walking wounded anyway," Glenn said, nodding his head slowly. "What can I do to get a cease-fire?"

I asked Glenn to elaborate on his father's objections to EDI. "Dad insists that it's a quick way to lose our independence. He says that like all chains, an interconnection of that sort would only be as strong as its weakest link. We'd end up basing our actions and decisions on what some, quote, 'half-wit is doing halfway around the world,' unquote."

"Sounds like your dad sees himself as a lonesome cowboy," I said with a smile.

"That's the understatement of the year."

"Do you suppose he subscribes to the notion that what you don't know can't hurt you?" I asked.

Glenn shook his head vigorously. "No, Dad's a darn sharp businessman. He didn't build a company like ours by hiding his head in the sand."

"Good. What you don't know *can* hurt you—and probably will. I'd rather know all about those weak links in the chain rather than get a nasty surprise when I could least afford it."

Glenn agreed but suggested that his father might also be worrying about the possibility that his operation could wind up being the one regarded as the weak link. "Dad has really done wonders over the years creating a powerhouse image for the company when in reality we are, frankly, marginal in some areas. He doesn't want the major suppliers to know that; while we get the job done, it can be touch and go." Glenn shrugged and added, "Probably he's a little paranoid that we'll lose an important line of products because we've got twenty delivery trucks instead of twenty-five."

As we talked, I explained to Glenn that he was going to have to do a sales job on his father to overcome the fear of the unknown and the technophobia that tends to increase during times of rapid change. I urged him to look for a problem that had occurred in the past that could have been solved with EDI.

"Christmas, four years ago," he said at once. "We do a big seasonal holiday business, and one item that's popular for about two weeks prior to Christmas is eggnog." He shook his head at the memory. "We routinely increase a standing order by about 5 percent to account for population growth in the marketing area, and that was the drill for the eggnog. We shipped it to the stores as part of the basic dairy-product delivery. A week later we went back in with another round of the stuff, figuring the previous shipment had moved out, but—most of it was still there unsold."

Glenn leaned forward and fiddled with a pile of books on the coffee table before continuing. "Not only did we have to take surplus eggnog back to the warehouses; there was more eggnog on the way from the suppliers."

"What happened to the retail demand?" I asked.

"Well, 'the Grinch who stole Christmas' wrote stories in the newspaper about the danger of high fat consumption during the holidays, and one of the products singled out was . . . "

"Eggnog."

"You've got it. With EDI, as I understand the technology, we would have known that eggnog sales were down almost immediately."

"That's right," I said. "The bar-code readers at the checkout counter would have given the EDI system a running tally of all eggnog sales. You could have seen the sales figures in real time, hour-to-hour."

"I know roughly how much a store should be selling on a daily basis by figuring the size and usual sales volume. For example, if Broadway Market is open from 9:00 A.M. to 9:00 P.M. and usually does twelve quarts a day, you don't have to be a rocket scientist to realize that by noon on Monday, if only one quart has moved, there is a problem."

I pointed out that an EDI system could also use a neural network system to learn and factor in an individual store's peak hours to accurately reflect the average sales. "There could be a big rush between five o'clock and six o'clock, but that would be accounted for based on past history," I said.

"Exactly. I could have even waited for the peak hours as a double check, and when nothing was moving out the door, taken a look at the other stores and notified the supplier to cut the size of the eggnog order."

"With EDI you wouldn't need to contact the supplier, since they would already know that demand had fallen off and would have adjusted production accordingly."

"More skim milk and low-fat cottage cheese," Glenn said.

"And ease off on the whipped cream," I said.

"Poor Tanya, it's too bad dairy farmers like her can't plug into EDI and automatically reprogram the cows to produce nonfat giblet gravy," Glenn said.

"Not gravy, but there are always industrial enzymes," I countered.

"What?"

"Farmers are manufacturing a product—but it doesn't just have to be food," I said. "Tanya's got a herd of four-legged chemical and pharmaceutical factories in her back pasture. But let's not get into that right now. I want Tanya to hear all about it and learn to play with her new cards.

"To get back to the eggnog. Consumers were probably substituting other products for the high-fat items, and EDI would have shown the changing demand patterns to allow you to react. Cranberry juice might have been the replacement beverage. You could have raised the price a few cents and recovered some of the losses on eggnog."

"Hell, if I had known what was going on, I would have set up special eggnog displays, cut the wholesale price, or done a coupon promotion," he said. "Anything to get that eggnog moving. It's perishable." He leaned back into the sofa and looked thoroughly frustrated. "Dad was climbing the walls. One of our smartest and biggest customers immediately saw what was happening and cut the price to the point that the eggnog was a loss leader. When the other grocery stores heard about that, they accused us of doing a deal—slashing the wholesale price for a volume order—and Dad ended up getting chewed out by all these irate dairy section managers from around the metropolitan area." Glenn paused and chuckled. "To this day he refuses to drink eggnog."

I excused myself a moment and went to find a pack of the new rule cards in the dining room. When I returned, Glenn was standing by the fireplace, examining a brass antique carriage clock on the mantel.

"I think you've got the best way to persuade your father to see the benefits of EDI. Eggnog and this card," I said.

"Health fads come and go. You may never have another problem with eggnog. But seasonal products—strawberries, turkeys, pumpkins—can all be affected by unexpected variables: the weather, a recession, labor troubles. If one of your stores' employees went on strike, EDI's sales figures would reflect that there is a problem, and you could adjust your orders early enough to avoid potential problems."

I gave him another example. "Red meat is not a seasonal item, but whenever there are stories in the press about hormones or chemicals or heart attacks, sales fluctuate. The demand for chicken and fish increases." I propped the card on the mantel beside the clock. "Wouldn't it be nice—and persuasive—if you could solve that future problem for your dad and yourself with EDI?"

Glenn picked up the card and nodded. "Yes, we wouldn't be playing catch-up all the time, and, you know, it would give us more control, not less. Dad spends a lot of his time trying to educate our suppliers about what the market will bear in the metropolitan area. The sales representatives for the major brands come in with high and unrealistic expectations, and he's the one who has to bring them back down to earth."

"And they probably think he is just being too cautious and conservative," I said. "If they saw the figures roll in on EDI, they would know exactly what the market is doing and realize that your company is not the weak link in the chain."

The Pot of Gold

Lonnie and Doug arrived, cutting short our discussion. Soon the others had joined us, and we were all gathered around the dining room table ready to begin.

"We're going to start off by having Glenn play against a phantom opponent," I announced and dealt out five cards, faceup. It was what is known in poker as a Big Cat: king high, eight low, no pair, which only loses to a flush (all cards of the same suit).

"Glenn's opponent is a national distributor that just bought an existing distribution operation in Glenn's hometown," I stipulated. "He's got money, an enormous infrastructure, executive talent, a solid reputation, and ambitious plans."

I picked up the deck to deal out Glenn's hand. "Ready?"

"I'm ready," he said, "but deal from the new tool and rule decks, if you don't mind."

"All right," somebody murmured in appreciation.

I dealt out the cards one at a time.

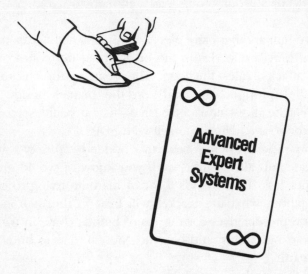

"Now, before we consider how you're going to play that card, I want Glenn to tell us why he is playing the game in the first place."

"Why?" Glenn asked.

"Playing for the sake of playing isn't enough," I said. "You've got to have your long-term objectives clearly defined. How else are you going to keep score? Without objectives, goals, targets, you could be losing and not even know it."

"Okay, I want to stay in business," Glenn said hesitantly.

"That's what I call minimalist goal setting. Shoot a little higher," I urged.

Glenn thought about it for a few moments. "I want to drive this new phantom distributor out of my market."

"Good. And . . . ?"

"Triple my sales volume and double my profits in ten years." Glenn nodded. "And to do that I'm going to have to add several new product lines each year and offer additional services to act as new profit centers."

"You're headed in the right direction, Glenn, but I'm still not seeing stars in your eyes. Dream a little. One of the most subversive ideas is the one that requires us to be 'practical' and 'levelheaded.' If you don't believe in rainbows, you'll never see the pot of gold. Describe the rainbow to us."

"Fantasy time," Maury Andrews grumbled.

"No, not fantasy—destiny. It's important to discover and then achieve one's own destiny," I said. "You can't do that if you are imprisoned within a narrow, cramped, constricting view of the future. When Glenn said that his objective was to stay in business, that told me that his view of the future isn't very different from the present. It was like saying that his objective in poker is to break even or control his losses.

"Dwelling on current reality reinforces sameness. Dwelling on future goals—your idealistic goals—reinforces action.

"There's an old saying that you must saddle your dreams before you ride them. But even before the saddle goes on, you need an eye for good horseflesh, for the stallion that can run with the wind. Otherwise you end up on the back of a mule plodding along from the same old thing to the same old thing."

"But what good is a goal if it is unrealistic?" Lonnie asked. "You just end up chasing after mirages."

"Good question . . . good word. What was unrealistic a hundred years, fifty years ago, is so realistic today as to be commonplace. In 1927, Lindbergh's flight from New York to Paris was regarded as a historic stunt with very little immediate practical value. There were perfectly good transatlantic ocean liners, thank you, to and from Europe. The early aviation pioneers were very unrealistic.

"In the nineteenth century, immigrants came to America because 'the streets were paved with gold.' They were very unrealistic. Without that dream they would have stayed in the ghettos and the villages and the debtors' prisons of the Old World.

"The faster the pace of change, the faster the changes in what constitutes the definition of being realistic or unrealistic. In many ways, any of the new tools of technology can act as the genie in the lamp that grants our wildest dreams."

"So all I have to do is rub this computer chip, and a guy in a turban will materialize and do my bidding?" Glenn asked.

"The best I can do is a lamp shade on my head. Is that close enough?" I inquired. There was general agreement that that wouldn't be necessary. But the quip loosened up Glenn enough to get him to stretch his imagination.

"My company is a medium-size regional distributor. I'd like to go national." He paused. "How am I doing?" I winked to let him know that he was doing just fine. "Okay, I'm not interested in controlling the top national brands. You become a hostage that way, but I want exclusive contracts with solid performers in certain niches like frozen microwavable entrées, canned soups, herb teas, frozen low-fat yogurt, and—ready for this one?—cat food."

"I like it. Are you done?"

"Let me see. I could manufacture my own brand of fruit drinks and springwater. It'll probably never happen, but it's fun to think about." Glenn stopped and pondered some more. "I guess that's about it."

I looked around the table. "I predict that all of you are going to have Glenn's problem. We've got to start giving ourselves permission to dream big dreams. Somewhere along the line we switched off the dream machine. It's time to crank it up again."

"Dan, it's hard to have sweet dreams when you go to bed every night wondering where you're going to get the money to pay the mortgage next month," Tanya said, looking grim.

"And that's why it is even more important to immediately jump-start the dream machine," I said. "You're going to have to plan your way out from under that mortgage, Tanya. Isn't that right?"

"I guess so."

"Well, the planning process that I spoke about the other night depends on dreams. You can't go ten years out into the future and plan back to the present without taking a wild leap of the imagination.

"Once we get our feet off the ground, strange and wonderful things happen. Linear thinking, to name one, goes by the boards. The intellectual mechanism that carries us step-by-step-by-step from problem to solution—which I might point out hasn't been working very well when it comes to Tanya's mortgage—is suspended.

"And that's wonderful! Now we're getting somewhere. Instead of linear thinking, a bunch of pictures start popping up in our heads. It may be a postcard from the south of France or a snippet of a home movie showing children playing on a green lawn.

"Images are the roots of imagination. Visualize the future. Visualize your goals. The old proverb says, 'Without vision the people perish.' Why? Because without a vision of, say, the promised land, a land of milk and honey—how's that for a picture?—the people remain lost in the desert. There is nothing to strive for, nothing to live for.

"John F. Kennedy knew the power of an image when he set land-ing on the moon as a national objective of our space program. He could have held out all the scientific and research benefits or empha-sized the military advantages. But he knew that to capture the imagi-nation of the American people and to energize them into action, he had to create a strong visual image of that first human footprint in the lunar dust.

"I tell my corporate clients to tear up their written mission state-ments, those massive unread tomes that sit in file cabinets. There should be a picture of the future—an image, a vision—in the lobby of the company's headquarters or the employee cafeteria; anyplace where it is visible instead of invisible. The executives, the entire work force, every customer, should be able to put themselves into that pic-ture."

I turned back to Glenn. "Give me a picture of your success in ten years." He frowned and started to furrow his brow. "No, don't think about it. Picture it. Quick!"

"I'm on a plane flying to meet a new customer."

"Good, you're obviously not just a local or regional distributor if you're on a plane," I commented.

"We're coming in for a landing, and I'm looking down at the Eiffel Tower."

"Congratulations, you're an international distributor."

"But that wasn't even one of my goals," Glenn objected.

"Well, why not?" I asked.

"Hmmm . . . good question."

Accumulating Intellectual Capital

"By visualizing the future, we now have an idea of where we want to go. Okay, Glenn, your first tool card was advanced expert systems. How do you play it?"

"First off, since I don't have any experience nationally or interna-tionally, advanced expert systems might give me a crash course in the things that I need to know: tariffs, taxes, agricultural regulations, trans-portation networks."

Brenda Jacobs raised her hand. "Am I wrong, but isn't Glenn going to have trouble finding an advanced expert system to exactly fit his requirements? You don't just walk into a software store and buy one."

"To keep development costs low, Glenn could go to a local university and talk to the dean of the computer science department. It's very likely that you will find students or perhaps faculty members who are ready, willing, and able to help you put together an advanced expert system that would fill the bill. To save additional time and money, start with what is called an expert system shell, a program that is already an expert system without the expertise. By identifying the experts and interviewing them in a special way, you will have created a usable system. Assembling the system would not only be a learning opportunity, but the project could attract grant money and licensing income by making the end product available to other industries."

"I'll bet that the United States Chamber of Commerce would jump at the chance to provide experts on trade and taxes and those kinds of things," Lonnie suggested. "Their members could all use the same kind of help that Glenn needs."

"The surface of this technology has barely been scratched," I said. "Anyone could use it. I can't think of an industry, a profession, or occupation that wouldn't benefit from advanced expert systems."

"Well then, why don't I really play the card, Dan, and develop an advanced expert system that I could sell to my customers for their use, which would tie them more closely to me now and in the future? It would be a service that my competition is not providing."

"You're onto something there, Glenn," Maury said before I could answer. "Instead of selling it to them, I would give it to them. That would make sure that all would get it. That would also make sure that all were tied into your company. The system could answer your customers' questions about new products, help them with marketing and in-store displays, and give them advice on accounting practices. They would be able to move more of your products through the retail channel by utilizing a level of expertise that only the largest chains can muster. Come to think of it, I think I should offer an expert system to my own customers."

"Let me tell you, the chains are one of our biggest headaches," Glenn said. "They play one distributor off against the other, squeezing our margins to the point that nobody makes a profit. Then on the retail end, they murder our little grocery store customers. This card would help on both counts and give me a chance to compete on a basis other than price."

"I could use an advanced expert system on taxes alone," Maury said. He and Glenn were talking past each other as each man got caught up in the excitement of customizing the technology. "Changes in the U.S. tax code are bad enough, but I've got twenty different countries to worry about."

Lonnie interrupted him with a question: "Let me see if I've got this straight. If I had an erosion control job that was giving me trouble, I could sit at my computer with one of these advanced expert systems up and running. I'd explain the situation, and a leading landscape design authority on erosion control would come on-line and advise me?"

"Yes, but the expert wouldn't be actually sitting at the end of the phone line or the fiber-optic network answering your questions 'live.' The advanced expert system is a flexible and interactive reference work," I explained. "The expert's knowledge base, including elements of his or her inductive and deductive processes, has been entered into the system, and the software matches your questions with the appropriate data.

"Actually," I continued, "the system could pose the questions, since many people aren't very good at asking the right ones, and the user would respond by selecting options. It would ask if the problem is water or wind erosion, for instance, and you'd use your computer's mouse to click on water. Then there'd be another series of questions: Is the site on a floodplain, adjacent to a water course, or subject to runoff from rainfall? An option would be selected, and another series of questions comes up. Does the problem occur in summer or winter? As you can see, the system moves from the general to the specific.

"The United States Navy was one of the pioneers in applying advanced expert systems. They wanted to teach young submarine officers how to skipper nuclear subs without waiting for those men to spend years at sea accumulating experience. Artificial intelligence experts came in and debriefed veteran commanders on every aspect of the job. They asked tens of thousands of questions. The goal was to capture the combined years of experience the commanders had so that other, less experienced people could benefit."

"It sounds like we're headed for submarines driven by robots," Samantha observed.

"No, what we're headed for are submarine commanders who

have been trained by the very best skippers in the navy, no matter whether they are still on active duty or unavailable from other assignments to serve as instructors. Think of the value of having John Paul Jones to consult on naval tactics, or Admiral Halsey."

"You mean advanced expert systems can call experts back from the dead?" Doug asked. "It sounds like technological voodoo."

"No, Halsey and Jones may rest in peace. But starting now we can stockpile a legacy of expertise for the future. If Stephen Hawking, the author of *A Brief History of Time,* participated in an advanced expert system, we would always have the benefit of his mental acuity. Remember, though, with rapid change expertise also becomes obsolete. John Paul Jones might not be a lot of help skippering a guided-missile frigate."

"Yes, Dan, but how many times have we all said 'Too bad old Joe isn't still around—he'd solve this problem in a minute'?" Maury asked. "I think we could use our best and brightest in-house talent—the R&D people, for instance—as a resident advanced expert system. When they retire or quit, the company still would have the benefit of their talent. We'd be accumulating intellectual capital."

I completely agreed with Maury. "Intellectual capital is usually squandered by large and small companies that otherwise closely guard more tangible assets like real estate and patents. There are exceptions, though," and I offered one: "IDS Financial Services, the financial planning subsidiary of American Express Company, recently tapped the expertise of its best account managers to create a version of an advanced expert system that it called Insight. IDS found that it could significantly improve the skills of its planners to the point that the worst of them exceeded the proficiency levels attained by planners who had an average success rate prior to the establishment of Insight. In a four-year period, the percentage of customers who closed their accounts dropped by more than 50 percent."

"What Doug said about technological voodoo might get to be a real problem with this kind of technology," Brenda said. "You end up running the hospital or business by consulting a system built around people who were on the job five or ten years before. I can hear management insisting that the advanced expert system just had to be right no matter what the real, living, breathing, working people had to say. After all, who do you think you are? The experts are geniuses in their field, or they wouldn't be part of an expensive computer program."

"That's a good point," I said. "Does anybody else have a response to Brenda?" I asked.

Tanya was the first to speak up. "She's right if they froze the advanced expert system in time and never changed it. But presumably they would be continually feeding in new information, developments, and expertise." She shrugged. "In farming there's lots of stuff that stays the same for decades. But take veterinary techniques for treating live-stock. If I was using an advanced expert system for that, it would be useless if it wasn't being continually revised."

"Keeping these systems up-to-date is simply not a problem," I said, picking up where Tanya left off. "There are two types of advanced expert systems. One of them amounts to a closed loop. For example, your Buick mechanic needs help with the 1992 Park Avenue transmission, and so GM provides an advanced expert system aimed specifically at that model's particular characteristics. After several months, the company's engineers have a pretty good idea of what makes that transmission tick. They can send out the software on a CD-ROM because it is probably not going to need updating; if it does, they can issue a second disk. The second kind of advanced expert system deals with situations that will always be in flux, such as the treatment of cancer, or, in Tanya's example, veterinary techniques. Therefore, it uses an on-line delivery service like satellite or a fiber-optic network.

"To respond to your concerns, Brenda, about a five- or ten-year-old system making decisions in the hospital, an on-line program is constantly being updated to reflect current medical and scientific reality. Obsolete expertise is being flushed out of the system. Furthermore, that system is not making decisions. Decisions are made by people. Management would be dead wrong to try to give an advanced expert system the final judgment call. Computers don't have judgment. Only human beings have the capacity to sort through the myriad variables involved in complex decision making. That's what we're good at.

"Computers, on the other hand, excel at quantitative analysis. With the latest storage capacity and processing power, new technology can handle vast amounts of information—slice it, dice it, chop it, shred it. Then comes the qualitative analysis, and the sign goes up: Men and Women Working!

"The advanced expert system can tell you, Brenda, that in July 1992 Doctor Martinson of San Diego General treated a stubborn strain of TB in a certain way. And it was a brilliant example of medical intuition. But it is up to your hospital's chief of internal medicine to say 'Martinson's work is very helpful—I never would have thought of that—but this case is slightly different.'"

Tools Not Tyranny

"From a manufacturing standpoint, thanks to advanced expert systems there's no need to keep re-inventing the wheel," I explained. "For example, Ford Motor Company tracks the equipment and processes used to make electronic components in all of its plants. An engineer who wants to change a production technique—let's say the angle at which a circuit board is soldered—can ask the system for advice and receive the information that when a plant in Brazil rotated its circuit boards by ninety degrees, the solder flowed at the new angle in such a way as to short out the circuits. Therefore, the engineer says, 'Whoops! Bad idea.' Or, 'I think I'll try it at sixty degrees.'

"Again, a key point to keep in mind here is that the engineer can still use his or her best judgment. The software doesn't supplant human control. The advanced expert system is another input. The engineer can accept or reject the advice. In the case of the circuit boards, the determination could be that solder with a different composition and viscosity wouldn't flow as freely as it had in Brazil. Technology is a tool—not a tyrant.

"Many people worry that computers are going to replace human beings. Samantha raised that possibility about submarine commanders. And you should worry—if. If you are one of those people who stops developing his or her unique human capabilities. Computer programs are being designed to replicate, as closely as possible, many human functions. An important human function is the ability to continually upgrade our mental database. We learn. We accumulate experience. When that process comes to a halt, however, the computer has the advantage because it goes right on updating its database.

"Human intuition, insight, and creativity depend on a well-stocked database. Twelve or sixteen years of formal education—no matter how good—isn't enough anymore if 'graduation' means that the database essentially stops expanding.

"That's why I have this card in the new rule deck." I held it up for everyone to see:

"The two must go together. Technology is nothing more than a tool. People are tool users. But if they don't have the necessary skills, then even the most sophisticated technology is only a blunt instrument.

"The shortsighted refusal to upgrade people—or to upgrade ourselves—is a crucial factor in sabotaging technological change. I wish I had a dollar for every time I've called a company and been told, 'I'm going to transfer you but I may cut you off . . . we have a new phone system.' Or, 'I didn't get the message because the new voice-mail setup is driving me crazy and I've quit checking it.'

"What happens next, Glenn?" I asked.

"Well, the customers start getting annoyed, and they take their business somewhere else to avoid the hassle. The bottom line goes soft, the boss panics and lays off workers, which means that fewer unskilled people are doing more work using the new technology. . . . "

"The new technology keeps going haywire," Lonnie said, jumping in to finish Glenn's thought. "More customers leave, more losses, more layoffs."

"It's a classic death spiral," Maury observed. He frowned and took off his eyeglasses. "Maybe that explains why communications in my company break down more today even though we have E-mail. I had consultants tell me that there were too many layers of middle management for effective communications. Now, middle managers are on the endangered species list, there's state-of-the-art E-mail, and we still don't know what the hell is going on." He glared at me. "I was getting ready to junk the E-mail system."

"Talk about death spirals," I said. "Add new technology at great expense but without upgrading the people who will use it. When the technology doesn't work right, dispose of the people. The problem persists, eliminate more people. No solution in sight, scrap the new technology.

"Viola! The worst of all possible worlds. And meanwhile," I added, "the competition has played the new rule card by upgrading technology and upgrading people."

Asset Appreciation

"Aren't we really talking about priorities?" Samantha inquired. "People come and go, but an investment in hardware is going to stay put. You can amortize the investment, in other words."

"There are some interesting figures on that," I said. "It varies from industry to industry, but Westinghouse claims that its average technology becomes obsolete in five to seven years. Wow! That's an awfully short amortization period to justify shortchanging your people in favor of hardware.

"US West, the telecommunications company, has two classes of technological investment: one, technology that creates knowledge; two, technology that automates support functions. The company says the return on knowledge is about twice what it gets from automation. And Motorola calculates a return of thirty-three dollars for every dollar it spends on employee training."

"Hold on a minute," Maury commanded. "I just looked it up the other day. The median return on assets for Fortune 500 companies in 1991 was 4.8 percent. You're telling me that Motorola gets a ratio of thirty-three to one on its employee training investment? I don't think I have a tangible asset that comes anywhere near that kind of performance. Certainly not our real estate holdings these days."

"Plus," I said, "tangible assets depreciate; they wear out. Factories get old and outmoded, machinery needs to be replaced. If upgraded, employees actually appreciate in value. I call it human capital. If your accountant can't come up with a way to measure human capital, you need to either immediately implement a new investment program to upgrade your people—or get a new accountant."

Experience Counts

Glenn was eager to resume playing the game. "Can I request a card, Dan?" he asked.

I chuckled. "You're learning. One of the things about the new game is that the cards are available to everyone. There really isn't a dealer. All you have to do is know that the new cards exist and you can deal yourself a winning hand." I pointed to the cards. "Which one would you like?"

"Object-oriented programming. The one that Maury got the other night."

I handed it over and asked Glenn what he was going to do with it.

"Object-oriented programming is the perfect way to invest in my people. I've got a total of about two hundred years of experience sitting out in my warehouses—speaking of human capital. My drivers and warehouse floor people could use object-oriented programming to develop new and flexible software programs to fit our special situation. I've never been happy using an off-the-shelf inventory control system; half the time it doesn't quite fit, and we wind up with the wrong products going to the wrong locations. We can't afford an in-house programmer to customize our software."

"Glenn was interested in EDI. Is there any reason he couldn't teach his people to use object-oriented programming to custom-design that system as it interfaces with his operation?" Tanya asked me.

"Absolutely no reason at all. Probably every individual in Glenn's organization would like EDI to do a specific job. The folks in the meat department need to know what grades of beef are coming out of the packing plant. If there is a new product being introduced, somebody else is going to want to know that it's coming so they can make warehouse space available. Each person could design his or her own part of the system."

"I see my three cards—EDI, object-oriented programming, and advanced expert systems—as real equalizers against my phantom opponent," Glenn said. "He is a big-time operator, and while I'm no ninety-pound weakling, I'm sort of a light middleweight. Yet, I plug into an EDI network and get the same access to information that he has."

Brenda was nodding her head enthusiastically. "And you get world-class advice from an advanced expert system. I'll bet you could obtain guidance from other distributors who have gone head-to-head with your opponent in the past. They may know his every move, his strengths and weaknesses."

"An advanced expert system is fine, but personally I think object-oriented programming is a real ace," Doug said. "Glenn gets the benefit of local people who know the local market setting up his system, and they get a kick out of learning new skills instead of spending eight hours a day driving a forklift or filling out inventory control sheets. Some of my former students work in Glenn's warehouse, and I'd love to see them go in that direction."

I reached for the new tool card deck. "Doug, by mentioning inventory control sheets, you gave me an excuse to deal out another card to Glenn."

I continued, "This card takes us beyond today's notebook computers by further shrinking the size and eliminating the keyboard. The electronic notepad uses a penlike input device. Doug's former students will conduct the warehouse inventory by using electronic notepads. The same goes for filling orders. There won't be any paper to contend with. The notebooks will be linked to the EDI system, which will allow the people in the warehouse to see that the truck being loaded for Fred's Maxi Market should get another case of orange juice beyond what the order calls for because the Boy Scouts' camp director just popped in unexpectedly and bought out the rest of the store's stock of juice. On the other end, the juice supplier will immediately see that Glenn's warehouse is going to need an extra case to replace the one that's going on the truck."

Glenn was eager to play with that card as well. "My drivers could use those electronic notepads right in the store as they count the stock on the shelves using a bar-code input device to save time. When they head back to the office, the data could be dumped into the system."

"Or, they could use a PCN network or digital cellular telephone technology or even a digital radio network to instantly transmit the data right from the store," I suggested. "The EDI link would kick in and tell the salesperson to offer the merchant a special price incentive."

I went back to the deck. "By using electronic notepads we've actually encountered another new tool card."

"How are you going to play this one, Glenn?" I asked.

"I'm not sure," he said, looking uncertain.

"Well, earlier this evening, before we got started, you were telling me about a Christmas eggnog problem that you had a few years ago. If your customers' shopping carts had all been equipped with flat-panel display screens, you could have done quick, inexpensive in-store promotions for eggnog that shoppers would have seen as they pushed those carts around the store. It might have helped counteract the effect of the holiday season fat-phobia articles in the newspaper."

"You mean each cart would have its own TV set?" Maury asked.

"I've heard of that, but I've never seen the application up and running," Glenn replied.

"Yes, each cart has its own display that's about half an inch thick, or less, and the programming changes depending on where the shopper is located in the store. If he or she is wheeling down the aisle that displays salad dressing, the screen could be showing a special on olive oil or introducing a new blue cheese dressing. Turn the corner and start past the coffee, the screen switches to a Maxwell House two-for-one deal."

"I don't distribute Maxwell House," Glenn shot back.

"Then make some money by selling them advertising time. The major-brand products are going to want access to the medium. You'd be entering into a relationship with Maxwell House without becoming dependent on that particular brand for a major portion of your revenue. Instead, the brand is dependent on you for a service it's not getting anywhere else."

"It would also be a solution to the problem Glenn has with the chains clobbering his smaller customers," Doug suggested. "Particularly if the screens also displayed floor plans of the store and comparative pricing information." He pulled his chair closer to the table. "Could the screens be interactive, like a computer terminal?"

"Sure, a computer is basically a bunch of chips and a screen. Touch a small icon that looks like an old-fashioned adding machine and a calculator would appear on the screen so that the shopper could keep track of what he or she is spending. Touch an icon shaped like a chicken and a special on poultry would be displayed along with directions on how to reach the appropriate section of the store."

"How about putting a bar-code reader in every cart along with the screen and automatically doing a running tally to save time at checkout?" Brenda asked.

"I'll do you one better," Lonnie said. "Make the cart a rolling checkout counter. Rig it up to hold the bags open. As the products go in, the price is read, the total gets rung up, it accepts a credit card and issues a receipt. When they're done the shoppers hand the receipts to a security guard at the door and proceed to wheel the groceries out to the car."

That started a technological bidding war. Before it was over, robotic carts were moving through the store seeking out the products on a shopping list that had been scanned into memory. On the bottom of the screen a reminder flashed on telling the shopper that he or she was low on sugar (based on the family's consumption rate and the last purchase two months before). The robotic carts were even offering security in the dark parking lot by providing a panic button to set off an alarm and a direct voice link to the store's guard post. When the shopper was safely in the car, the cart headed off to a holding area to wait for the next shopper. Brenda was about to raise the stakes even higher by suggesting that "the cart could be eliminated altogether if shoppers were equipped with hand-held bar-code readers that would zap the required products and flash the order to the back room to be filled and sent—"

"Hold it!" Glenn said, cutting her off in mid-brainstorm. "Let me learn how to walk before I run. This advanced video display technology is pretty expensive in its own right. I'd probably break the bank if I went overboard."

"The cost of flat-panel video displays is dropping at the rate of 30 to 40 percent a year," I said. "You could ease into the technology by initially introducing a few larger screens positioned at key points in the stores of your larger customers. Those displays could do double duty as training tools for the store's staff by broadcasting seminars on customer relations, antishoplifting techniques, spoilage control, and that sort of thing. The training could be an additional revenue generator for your company to underwrite the new technology."

"I could also use the flat-panel displays to introduce new products simultaneously to all my store managers and owners without having to spend hours making individual presentations at each store," Glenn said.

Samantha stood up to stretch her legs. It was a subtle reminder that we had been at the game for more than three hours. "Since one of Glenn's goals is to become a national and international distributor, why not distribute training and technology to the food industry?" she asked. "It seems like a natural fit. If independent food retailers are going to survive chain competition, they need somebody like Glenn who knows their business and can point them in the right direction."

"And it doesn't mean that I have to get out of food distribution, either," Glenn added in response. "I can do both. The technology would just be another product."

"There you go," I said. "We all need to determine what business it is that we are really in. Tunnel vision could persuade Glenn that food and grocery items are the be-all-and-end-all. Now, though, he is beginning to understand that he and his father have created a distribution channel that can be used for other purposes and products.

"Many hoteliers, for another example, are realizing that they are not just in the business of renting out rooms overnight, but that they are offering corporate support services. Another client of mine has a vast natural gas pipeline network. In the late 1980s, the question was how to increase revenue during a recession and an energy glut. The answer was to get to the heart of what that pipeline system was all about. And what is it all about? It's a delivery system. Today, a fiber-optic network is running through that old pipeline, earning far more than the natural gas, which he is still supplying. The company beat the down cycle by using new technology and new thinking.

"Now, let's go back—"

"Dan, explain something before we get too far off track," Doug requested. "When we were talking about the robotic shopping carts, you mentioned that it could remind the shopper to buy sugar based on prior purchases. What technology is involved there?"

I reached for a card.

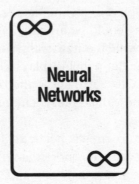

"Neural network technology allows the system to learn from past experience. If the shopper had a credit card or an ID number, the neural network remembers that there have been regular purchases of five pounds of sugar every two months. Hence, it will issue a reminder about the sugar if two months go by without a purchase. A neural network would also notify Glenn that one of his customers did not place his usual Monday order for a hundred cases of laundry detergent, tipping him off to the possibility that there's been a mistake—somebody forgot the order—or that business is off at the customer's store. Another possibility is that a competitor has given the customer a better deal, and if Glenn follows up on the warning he has received from the neural network, he can fight back to save the account." I glanced at Doug to see if he was satisfied. The teacher was beginning to look a little numb from all the new information. The others were in the same condition, so I moved to wrap things up.

"Let's go back and figure out where Glenn stands against the Big Cat, the poker hand held by his phantom opponent. Glenn, your market has expanded instead of contracting; your customers are stronger, you're operating nationally and internationally; you're no longer competing exclusively on slim price margins; and you're part of an EDI network that allows you to react quickly, plan ahead, and cut inventory costs. Anything else?"

"My work force is more productive because of the object-oriented programming and electronic notepads," Glenn immediately answered. "The advanced expert system keeps me ahead of the industry trends, is a major income source from my customers, and gives me the information I need to establish my fruit drink and springwater product lines."

"Who wins the hand?" I asked.

"I do."

6

From the Edge of Disaster to the Cutting Edge

When to Hold 'Em

"Tanya, it's your turn," I announced as the players took their places around the dining room table for another evening of learning how to play with the new tool and rule cards. "Are you ready?"

"Hardly. I spent the last week trying to come up with a vision of the future, and all I got was a mental picture of a suburban subdivision," she said with a rueful smile. "Something tells me I have an attitude problem."

"You're not unusual," I said. "When it comes to a futureview, that all-important set of expectations about the future, pessimism is often the farmer's favorite crop."

"It's certainly not a cash crop," Tanya said. "But at least we're not harvesting the grief we did back in the mid-1980s when I inherited the farm from my mother and father. We had a bumper crop of high interest rates and energy costs."

I nodded. "The pressure is off somewhat. What are you doing to take advantage of the respite?"

"I've been paying down some of my debt, putting cash aside, and expanding the herd a little bit."

"Sounds like you're being very cautious," I said.

"Better believe it."

"Cautiousness works if tomorrow is like today—or yesterday. What if it isn't?" Tanya hesitated before answering. "Go ahead; let 'er rip," I said to encourage her.

"It's going to sound like I haven't been paying attention to the new tools and rules cards but . . . if tomorrow isn't like today, I guess I'll just have to cross that bridge when I come to it." There was a hint of defiance and stubbornness in her eyes.

I rose from the table without responding, walked over to the flip chart, and wrote out an old English proverb.

The best throw of the dice is to throw them away

I went back to my chair. "Tanya is the one real, dyed-in-the-wool gambler in this room. She literally is willing to bet the farm every year on a throw of the dice. The weather alone—and isn't that the ultimate variable?—makes farming a huge gamble. There's no way to control the rain, or the sun, or the wind. The gambler-farmer sows the seeds and hopes for the best."

"Maybe we're fatalists, Dan, rather than hard-core gamblers," Tanya suggested. "There's only so much you can do, and then it's time to sit back and wait for nature to do its thing."

"Gamblers, just like farmers, use their skill and intuition, then sit back to wait for the odds to do their 'thing,'" I said. "But farmers, since humankind harvested the first crop, have usually only had to gamble against a single opponent—nature. They could win a few and lose a few and still survive. Today, farmers are playing against nature, against the commodity markets and the banks, against the government, against international competition, against wholesalers, retailers, and the consumer. It's really farmers against the world, and as such, it is a no-win situation. Little wonder that so many gambler-farmers decide there's no point in 'holding them.' In ever increasing numbers, they throw in their hands. They sell out."

"Come work for me, Tanya," Lonnie said. "Landscaping and farming have a lot in common. We both make things grow, and lawn cows are a lot less trouble than the real thing . . . no manure pile."

"Don't you dare," Samantha warned. She reached for the deck of tool cards. "Play this one." She threw out the EDI card. "Plug into Glenn's EDI network, and that will help even the odds."

"How can I afford to do that?" Tanya asked. There was exasperation in her voice. "That's going to take money, and the last thing in the world I plan on is going deeper into debt. Now that the credit crunch is over, I must get a call a week from somebody who wants to lend me money. 'Don't you need a new tractor, dear?' 'How about a state-of-the-art milking machine, honey?'"

"That's exactly what Dan is talking about," Maury Andrews said. "They want you to gamble that you'll win big enough this season or next to pay off the loan and have a little left over beside."

"And if I don't?"

"Borrow some more."

"Until they foreclose."

Maury nodded.

"Well, I'm not playing."

I pointed to the flip chart. "Wrong! Throw the dice away—but don't fold, don't die from attrition, don't sell out—pick up the new cards. Hold 'em—and hold your most valuable assets: experience, equipment, and land. Playing is not optional, Tanya. You're going to have to play to survive. Just continuing to roll the old dice or rely on the old cards isn't enough. And, to use another gambling term, there's no standing pat. Those who stand pat stand to lose.

"Recently, I was in the South making a series of speeches; it was Atlanta, I think. I was in a cab from the airport talking to the driver about the breakup of the Soviet Union and other dramatic changes that had taken place. He said to me, 'Boy, the next twenty years are sure going to be interesting to watch.'

"He actually thought that he was going to be able to sit back and enjoy the show—no way! If he does—if any of us do—the show is not going to have a happy ending."

"I understand that, Dan, but how do I play if I can't pay?" Tanya asked.

"First, define your *ideal* goals. This will give you direction and a positive vision of the future. Second, learn and apply the cards—including this one." I dealt from the rule deck.

"And third, you have to get excited about your lack of knowledge and challenge yourself to find out more."

"Stop right there, Dan," Brenda requested. "If she lacks knowledge, how can she creatively apply technology?"

"Don't edit out the most important part—'and challenge yourself to find out more.' We can't afford to be afraid of things we don't understand. Think about it. You will be introduced to so much new technology in your future, you will find that you have a lot of 'lack of knowledge.' There *is* a benefit to having a lack of knowledge. You can more easily come up with creative applications for something that is new. Experience often blocks us when we try to be creative. We need to blend our past experience and expertise with an open mind toward learning about new things." Lonnie was rapidly taking notes, so I paused to give him a chance to catch up. "Often we fear what we don't understand, and we run for cover.

"By getting excited about our lack of knowledge—and boy, is there a lack of knowledge about how technology is changing our world—we have the option to rejuvenate ourselves.

"Take EDI, for example. Until a few nights ago, most of you had only a passing acquaintance with EDI. Now, Samantha is able to suggest it as a card to Tanya. But Tanya says it is too expensive and rules it out. Discussion closed." I held up the latest rule card. "What if she knows less about EDI than she thinks? Let's challenge Tanya to get excited about her lack of knowledge and to find out more. And then it will be time to creatively apply technology.

"Start inside and work outward. Developing the core technologies and the enabling technologies for EDI was very costly; Tanya's right about that. Somebody else footed the bill, though. Today, Tanya, you can have a desktop personal computer with the processing power of a supercomputer that only a few megacorporations could have afforded in the early 1970s. Semiconductor memory back then cost roughly five cents per unit or bit. By 1990, it was one-thousandth of a cent per bit.

"In the 1950s, the world was in awe of the UNIVAC I computer. It was a monster, weighing in at 16,000 pounds with 5,000 vacuum tubes. And what speed! A thousand calculations a second. Today, the equivalent computer is about the size of the office watercooler and performs 55 million calculations a second. All that power and speed is available to you at a fraction of the original cost. Yet, it is worthless if you do not apply that technology in a way that serves your needs and your goals."

"Well, I don't need 55 million calculations a second to milk a herd of cows."

"How do you know? You can't even tell me what you want the future to look like in ten years."

Hit the Road

Tanya was stumped. She knew what needed to be done, but couldn't quite cross the threshold and start dreaming.

"How about combining cows and some crops that aren't as much of a glut on the market as dairy products," Doug gingerly suggested.

Tanya pondered the idea. "Like what?"

Doug shrugged.

Samantha gave it another try. "Plug into Glenn's EDI and find out. It seems to me that farmers have dropped out of the real world. You can plug back in with EDI and the other technologies and find out what people are buying."

"The problem is still cost," Tanya replied. "EDI is perfect if you're a major retailer with all kinds of outlets or a gigantic industry."

"Let's pursue that by using the last rule card I dealt, *Creatively apply technology,* by making use of a device that has been around since it was invented in the nineteenth century—the telephone," I said. "When we discuss EDI, we start thinking about fiber optics, and parallel processing, and direct broadcast satellites; it's all high-end technology. But why not use the low end? Use it for the purpose that EDI was originally intended for—to gather paperless information from one end of the manufacturing and retailing chain to the other.

"Tanya obviously doesn't have the resources of a Gap or a Wal-Mart. She also doesn't need the vast quantities of data that they require on a minute-to-minute basis. Hence, she doesn't need the costly high-end capabilities. Instead of saying 'EDI is not for me,' modify the technology and, in the process, make it economically feasible."

"With the telephone?" Lonnie asked, uncertain that he had heard correctly.

"Tanya doesn't need a high-bandwidth information delivery system. The copper wires that the telephone company ran into her farmhouse will do just fine," I said. "An E-mailbox that is set up to automatically receive the information that Tanya needs to know would serve her just as well as a full-bore EDI system.

"Think of it as a system of roads. EDI is an interstate superhighway whizzing traffic from point A to point B to points C, D, E, and F. But what about points G, H, and I? Those places are connected by state highways. Points J, K, and L are linked by secondary county and town roads."

The analogy registered with Glenn. "I need to know about hundreds of different products, about production and transportation bottlenecks, and about consumer buying habits—all of that comes to me via EDI, the interstate superhighway," he said. "Tanya, on the other hand, and every dairy farmer in the state for that matter, would suffocate under all that information. But they do need to know about consumer demand for milk, a change in the federal milk price supports, and a new program to promote an increase in dietary calcium that is boosting the sale of skim milk. My version of EDI could be programmed to flag that information and send it to Tanya's E-mailbox, in effect using a two-lane road rather than eight lanes."

"That's assuming she has an E-mailbox," Maury Andrews said. "Our E-mail system is just for management."

I picked up the rule deck and dealt Maury a card.

"We're going to have to rethink our traditional internal and external business relationships," I said.

"Why? It's the C-word again—change. Let me introduce a little history of technology. Just a little," I promised.

"Starting about two hundred years ago, we put machinery in our workshops and turned them into 'factories,' a term borrowed from the French to describe a totally new working environment. Viola! The Industrial Revolution. But the revolution didn't spread everywhere, and I don't mean Europe versus Africa or the Middle East. It didn't even spread upstairs. The front office stayed pretty much the same, even though the bosses were running vastly different enterprises. Water power and the steam engine weren't much help in balancing the ledgers.

"Sure, the typewriter and telephone and calculator came along, yet there was nothing to match the assembly line and automation systems of the modern manufacturing plant. Nonetheless, the scale was so much larger than the old cottage industries, and the need for information—always the key consideration for any commercial activity—grew by enormous proportions. As a result, out of sheer necessity, management as we know it today came into being and got the job of managing this mountain of information, basically collecting and manipulating it by hand, one factoid at a time.

"The task was labor intensive. Without today's power tools, it made sense to have compartmentalized, hierarchical organizations. The lines of authority and communication ran vertically, not horizontally. The network, for all practical purposes, existed to shuttle information upward to where decisions would be made. In many ways, it was like an old-fashioned fireman's bucket brigade.

"In a big organization there's a lot of important information moving from the bottom to the top. If I'm going to keep my job, with its fat paycheck and fancy title, I can't worry about my colleagues to the right and to the left, the ones who are on the same horizontal plane as I am. Survival means pushing my pieces of information upward, adding value to them, and forgetting about what those other people are doing!

"One of the reasons the recession of the early 1990s was harder on middle management than blue-collar workers was that it finally dawned on corporate decision makers that technology now does a much better job of collecting and manipulating information than a cadre of $60,000- to $100,000-a-year managers."

Inclusion, Not Exclusion

"If I understand what you're saying, Dan, we first automated the factory floor, and now we're finally automating the front office," Brenda said. "Why did it take so long?"

"Remember what I said about the new technology cards—if the card exists it is going to be played. The cards to build semiautomated assembly lines, robotic welders, and other aspects of factory automation have been in existence for many years, and they were played. On the other hand, the awesome processing speed and power of today's computers have not been available, especially at a relatively low price, until recently, and that computing power is just now being introduced into the game in a major way; and we haven't seen anything yet!

"In 1971 a computer chip that held 1,000 bits of RAM, random access memory, was state-of-the-art. A billion-bit chip is just around the corner. Is that card going to be played?"

The question hung there unanswered until Doug spoke up. "The children and grandchildren of the blue-collar working class were encouraged to get educations so they could sit at desks and push papers to avoid the uncertainty and hard knocks that their parents suffered. It sounds like they jumped from the frying pan into the fire."

"Indeed they have, if we continue to use a pattern of organization—namely the vertical hierarchy—that excludes rather than includes. Indeed they have, if Maury continues to use an E-mail system and other high technology to exclude rather than include." I held up the *network with all* card and turned toward the CEO, and caught him looking a little sheepish. "The walls that we have used to compartmentalize and control information are being torn down by technology. A white-collar work force that is trained and deployed to utilize these walls and vertical lines and compartments is headed for the land beyond Oz where lamplighters and keypunch operators are fully employed."

"Dan, you said that external business relationships would also change. Would you address that issue for a moment?" Glenn asked.

"Yeah, and stop picking on me," Maury said with a laugh.

The Very Model of the (Un)Modern Major-General

"I may be getting over my head here, since anthropology isn't my field, but hierarchical business organizations are probably the result of entrepreneurs during the Industrial Revolution who suddenly had immense establishments to run—farflung empires in some cases—and who said to themselves 'How am I going to manage this monster?' What they did was to think about the only other form of human endeavor of the same magnitude—making war.

"Business and warfare have a lot in common: vast numbers of people, complex logistics, communication difficulties, and the need for quick decision making. Not to mention clearly defined objectives, specialized functions and responsibilities, and—most importantly—an enemy.

"For want of a better model, hierarchical military organizations and attitudes fit the bill. Pure Gilbert and Sullivan: 'The very model of the modern major-general' takes command of capitalism. And it worked, until now.

"Collaboration rather than confrontation is going to be the norm in the twenty-first century, and with forward-thinking organizations it already is. Once again, technology is changing reality. It always has and it always will.

"The good guys versus the bad guys, the us-versus-them mentality fits an era when *them* controlled a vital market or an industrial sector and were ready to battle with *us* for trespassing. The technological, resource, and market options were relatively limited. Four steam-powered looms in a single county in the north of England were three too many. One business's loss was another's gain. Now, with everything in flux, to paraphrase the old Walt Kelly line about meeting the enemy— we have met *them,* and they are *us.*"

First-Class Mail

Lonnie leaned back in his chair and nodded enthusiastically. "I think that us-versus-them ruined my old milling machine company. We even treated the customer like them, like the enemy. We didn't want them to know what we were doing and could care less about what they were up to. It was like trying to make love in the dark without touching or talking to your partner."

"A new variation on the phrase 'love is blind,'" I said. "Love is total sensory deprivation."

"And a new meaning to the question 'Was it good for you?'" Samantha added.

"That lowered the level," I said. "But in fact we can't make love or do business without knowing the answer to questions like that. If I were you, Maury, I would junk that E-mail system if it only went to management. Expand it to include all your employees, your suppliers, your distributors, and your customers."

"Even prospective customers," Glenn suggested.

"Exactly. Once the system is set up, it's extremely cost-effective to add mailboxes. The potential return is significant enough to justify giving the box free to a hot prospect or to a small supplier." I held up the *network with all* card. "Give your United States senators their own E-mailboxes and keep them up-to-date with information that they need to do their jobs. Where else are they going to get it?" I asked.

"I fax letters to my senators when I get irritated by what's going on in Washington," Glenn said. "And come to think of it, a fax machine, or better yet, computers equipped with fax cards could function like E-mailboxes by automatically transmitting data to designated faxes. It would just be a matter of tagging the data to determine who gets what."

"It would be a great way to let parents know what homework assignments the kids have been given," Doug said.

"Or reminding clients to water their rosebushes during a drought," was Lonnie's contribution.

My players were calling and raising each other like real card sharks.

Applied Technology

The exchange finally worked up enough momentum to carry Tanya over the threshold and into the future.

"I could probably get John Deere to set me up with an E-mail or voice-mailbox. Their dealers sure bug me enough about buying new equipment," she said. "My dairy co-op could use a system to link up all the members, and for that matter the Farm Bureau could too. Maybe start by using fax cards." Tanya drummed her fingers on the tabletop. "Okay, Dan, you wanted a futureview, so here goes: I'm going to start to agitate for E-mail at the co-op, and that's going to get the good old boys all riled up. They're still not accustomed to having me around. If it works the way you say it will, I'll go after the Farm Bureau, which I always wanted to get active in but I never seem to find a good time to start. And it's only E-mail that I'll be advocating; I'm not proposing that we colonize the planet Mars.

"I'll do what Glenn just thought of and give every politician in the state an E-mailbox. You think farmers wouldn't buy into an E-mail system that gave them instant access to all those polls?" she asked, certain of the answer.

"First E-mail, and then before they know it, neural networks," Brenda said in encouragement.

"Yeah, the high priestess of high tech," Tanya replied with a laugh. "Ms. Wizard."

"And then?" I asked.

"Good question. Don't they say information is power? I'll be sitting at the E-mail information hub. Give me ten years and I'll be running the Farm Bureau."

An hour before, the future had been a massive door with a huge padlock and sign announcing "Do Not Enter." Tanya, using just a few of the new rules and new tools as keys, had thrown it wide open.

I smiled and shuffled the deck.

Designer Genes

"If you're going to become the high priestess of high technology, Tanya, you're going to have to play with a few more tool cards," I said, "including this one."

"*Farmer Frankenstein's Daughter* or *The Holstein That Ate Manhattan*," Lonnie wisecracked when he saw the card.

"Sometimes I think they shot the film *The Fly* in my barnyard," Tanya said.

"You've both just identified the problem we have with this kind of technology," I said. "Films, and the mass media in general, have had a marvelous time scaring the wits out of people about the so-called dark side of science."

"It all goes back to what you were saying about people being afraid of what they don't understand," Doug said. "Average folks don't grasp this stuff all that well, and it makes them nervous. Now that the Soviet Union is no more and Russia is just another Third World country, we're going to need something to get us all worked up and anxious. Science is perfect for the new, improved bogeyman."

"Sci-fi and horror movies are already so popular, Doug, you'll have standing room only in high school biology," Samantha predicted.

"I wish it," he said.

"Don't wish, make it happen," I said. "Offer the kids a hybrid show, not a horror show. And give them a happy ending instead of a scare. Tanya could be one of your regular guest lecturers. Maybe she would bring in one of the minicows that she has introduced to her herd and the herds of other dairy farmers."

Tanya gave me strange look, but I continued. "More than twenty years ago a geneticist by the name of José Manuel Berruecos began an experiment in Mexico with thirty head of Brahman cattle. Only the smallest cattle were allowed to mate, and the others were slaughtered. After five generations of breeding he had twelve minicalves and eighteen adult animals that were about 2½ feet tall and weighed approximately 330 pounds. Brahmans are normally more than double that height and at least four times as heavy."

"The perfect house pet," Brenda said.

"I don't know about that, but I bet Tanya can tell us one advantage of having a smaller cow in her barn and pastures."

"There's less of her to feed," she said, right on cue.

"Ten minicows can graze on the same amount of grass as one normal-size cow," I said. Tanya's jaw dropped in disbelief.

"It's true," Maury added. "I remember reading an article that included pictures of the minicows a few weeks ago. I can't seem to remember all of the statistics, but I have a feeling Dan can help us out on that."

"There are a few more pieces of information I think you'll find interesting," I said. "While there are fewer pounds of pure beef per cow, there's more beef per acre. The minicow yields about four quarts of milk a day, one or two less than a normal cow. An extra benefit is that the smaller cows, according to Berruecos, are more docile."

"So much for the holstein that ate Manhattan," Lonnie said.

"And good riddance. When I identified José Manuel Berruecos as a geneticist, Lonnie and everyone else here said 'Genes—what do I know about them? Answer: genes mutate.' Ergo, the holstein that ate Manhattan, or for the real scare—C-A-N-C-E-R.

"We have learned how to read the genetic code of plants and animals. As a consequence mapping, restructuring, and remodeling the code to eliminate or enhance specific traits are possible. It has little to do with mutating genes, and a lot to do with the plain and simple breeding of those Brahman cattle to come up with a minicow.

"The guy in Mexico is not a mad scientist. He is using a technique that ranchers and farmers have employed for generations to improve their livestock. He is just doing it in a more systematic fashion. The same goes for all the other geneticists who, as of 1992, were cultivating an estimated 200 different types of genetically engineered plants. Many of those plants, like the minicows, are the results of basic crossbreeding; others are more complex examples of our growing skill at actually altering the gene code with advanced biological and biochemical tools."

Good Breeding

Glenn frowned. "I don't think I understand the difference between recombination DNA and crossbreeding," he said.

First I corrected what is a common mistake. "It's *recombinant* DNA, or rDNA. You could spend a lifetime studying this technology, but all you really need to know is that in the nuclei of cells there are chains, or sequences, of deoxyribonucleic acid (DNA) that act as the repositories of inherited characteristics.

"Say deoxyribonucleic acid as fast as you can," I demanded. Everyone tried, but Doug was the only one who managed it without stumbling.

"Now you know why we call it DNA," I said. "The whole idea of heredity hinges on the fact that we all have parents. If Mom and Dad are from a variety of tomato plants with red, oblong fruit that matures sixty-five days after germination, then their offspring inherit those characteristics via its DNA. With this tool, rDNA, we can say 'Yes, I like red, oblong fruit and the sixty-five-day maturation period, but I hate it when the tomato starts to rot two days after picking.' So, I take a bit of DNA from another plant—perhaps it's a summer squash like zucchini that can sit around for weeks without turning to mush—and add it to the tomato's DNA. Now, we don't have to pick the tomato when it's hard and tasteless."

"But then the tomato might look like a zucchini," Samantha said.

"No, because we haven't touched that part of the zucchini's DNA sequence."

"Sounds simple," Glenn observed.

"Believe me it isn't simple. Successful rDNA restructuring can take years of experimentation. But traditional plant breeding is even more time-consuming and uncertain. By initiating a total gene transfer between two strains of plants, which is what happens with crossbreeding, the hybrid offspring inherits genes conferring desired beneficial traits as well as undesirable genes that can affect crop quality, yield, or adaptability."

"Can't you also inadvertently breed out a beneficial characteristic, like resistance to certain diseases and pests, which doesn't show up right away, and suddenly there's a massive crop failure?" Doug asked.

"That's happened. In 1970 there was a huge corn blight caused by a single disease organism that attacked 90 percent of the United States' corn varieties. With today's new tools this scale of crop devastation doesn't have to happen. By using rDNA on two totally unrelated parent plants, researchers can actually increase the gene pool of desirable traits. It is even possible to go beyond that by combining traits of both plant and nonplant species."

"You mean it's now possible to cross animals and vegetables?" Lonnie asked.

"Don't be so shocked. But before we talk about that, let me finish with tomatoes. There's a biotechnology company in California, Calgene, that has added a gene to tomatoes that blocks the enzyme causing the fruit to rot, and they say they are getting an extra ten days on other varieties before spoilage sets in."

"I think I know where my customers are going to want to get their tomatoes from," Glenn said. "In the winter, their produce managers are constantly complaining about inedible tomatoes. Tomato sales just fall apart."

I pointed to the deck of rule cards on the table. "Isn't that a classic example of using technology to solve your customers' future problems? For that matter, it's a problem today and a problem tomorrow—declining sales of tomatoes because they taste like chunks of plastic."

"Maybe I should give up on cows, even minicows, and plant tomatoes," Tanya said.

"Maybe we should all buy stock in Calgene," Maury added.

"Calgene is just one of many biotechnology companies, and I'm certainly not here to recommend investments for your stock portfolio,"

I said. "Even so, I will give you a little overall financial advice. Biotechnology is very volatile, primarily because investors have been using traditional methods of assessing risk and reward. And that's a mistake. You can't expect quick payoffs. This research takes enormous capital, trial and error, a willingness to fail and keep trying, and the courage to get out in front of the competition even though it means risking a few arrows in your back. In the long run, biotechnology has great investment potential, and some of the biggest players in the financial game know it.

"In 1992, biotechnology was a $4 billion industry. The federal government is projecting it to expand to a $50 billion industry by the end of the decade.

"Population growth alone is enough to make that forecast seem conservative. The world's population is probably going to double in the next forty years. We'll have to feed all those extra people using existing agricultural land. Bulldozing the Amazon rain forest or clear-cutting in Siberia aren't the answers for obvious ecological reasons.

"The choice is mass starvation or innovation—biotechnological innovation. With rDNA applications there will be crops that are engineered to grow abundantly in specific places, like one type of cassava that flourishes in the heat and drought of central Africa and another type of cassava that can adapt to the far different conditions of the Indian subcontinent; crops that have higher nutritional value; crops that resist weeds, pests, and disease without chemicals.

"From a pure business standpoint, rDNA is an outstanding asset. With a pair of 'molecular scissors' it can be customized, patented, and used like a piece of property to generate royalties and licensing fees. As property, or a product, rDNA doesn't require vast warehouse space and complicated logistics. Labor, shipping, and energy costs are minimal; scarce raw materials, expensive feedstocks, crazy weather patterns—no problem."

"Hold on, Dan. I can hardly handle a pair of pinking shears, and you want me to use molecular scissors?" Tanya asked plaintively, shaking her head. "I'm no dummy, but this stuff is way beyond anything that I'm experienced with."

I went to the flip chart and wrote this word:

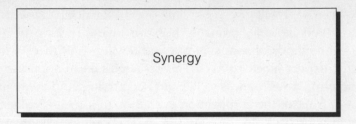

Synergy

"Muscles and nerves have synergy. They work together. The sum is greater than the parts. Scientists and farmers have synergy, but most of them don't know it. The lab is the lab, and the farm is the farm; never the twain shall meet.

"I think we demonstrated that farmers need the expertise of scientists to survive in the twenty-first century, just as they needed the expertise of metallurgists and engineers to survive the nineteenth and twentieth centuries. Where did the plow and the combine come from? Metallurgists and engineers.

"Now, let's reverse the dependency. I'm a scientist, and I've figured out a way to use rDNA to turn cows into living drug factories. In her milk, Bossy could also be producing something known as TPA, tissue plasminogen activator, a very important and very expensive blood clot–dissolving drug. But I don't know the first thing about cows. I can barely deal with laboratory animals." I pointed at Tanya. "I need this woman. Without her, and other innovative farmers, there's no possibility of field-testing and translating promising lab procedures into practical applications.

"I mentioned Calgene a little while ago. It relies on a few California farmers—farmers who know that the future is going to be far different from the present—to use their agricultural land and skill to grow things like a tomato that doesn't rot in order to see whether they come up with a commercially viable product. The farmers act as midwives to the scientists. They don't need to know how to use molecular scissors or any other biotechnology techniques.

"TPA, the drug that dissolves blood clots, costs about $2,000 a dose. Would you like a piece of that action, Tanya? Or is there something sacred about a glass of milk?"

"I never drink the stuff myself," Tanya said. "What's sacred is keeping the farm running and doing what I like to do." She paused. "You mean, I can get drugs out of that herd of mine?"

"If you recognize the new big picture and how it will affect you.

Example: A midwestern farmer who came to hear one of my speeches about technology wrote me a few years later to say that he still had his livestock and his barns and acreage, but that he wasn't a farmer anymore. He was in the biology business. That guy had just closed a deal to sell a herd of cattle from prizewinning bloodstock to an operation in Europe. He said he would have never thought of doing business internationally before, but he changed the way he thinks. He made a nice profit by shipping the entire herd before the cattle were born, in embryo form, in the placenta of a living rabbit. And, as he said in his letter, 'I saved tons of money on shipping.' He even got the rabbit back alive and healthy."

"How am I going to manipulate cattle embryos?" Tanya demanded.

"You're not, Tanya. But a twenty-four-year-old graduate student from the university up the road is going to do it for you. He'll probably work part-time for about ten dollars an hour to help pay his tuition.

"Funny thing about the young. They've got information—and it tends to be cutting-edge information—that their seniors don't have. How often do we have to ask our kids how to operate the computer or the VCR? They are in the process of becoming experts on the future. Why? Because of our age differences they know it's very likely they will spend many more years in the future than we will. They're not hung up on the good old days because they haven't had theirs yet!"

"Not unless we betray them by providing an education that's a generation or two out of date," Doug commented.

"Isn't that one of the reasons why teachers have so much trouble motivating their students?" I asked. "In the eyes of the students the curriculum doesn't seem to have any relevance to the future that they perceive they will be living in. And, in many ways it doesn't. In a way, we are asking our kids to become experts on the 1930s or the nineteenth century, and their youthful instinct tells them that it's a waste of time. I'm not saying that the study of history is a waste of time. I am saying that teachers often fail to show the future relevance in what they teach. Students will rebel or, worse, tune out and drop out. What a waste!

"But adults also tune out and drop out by refusing to recognize the changes that are taking place around them. We have to observe this rule," I said, dealing out a new card.

"Tanya is an expert in nineteenth- and twentieth-century agriculture. Now, she has to re-become an expert on the twenty-first century. Seems only logical, doesn't it? At her age, she is going to be living more than half of her life in the twenty-first century. She's got to update her mental database. Isn't that one way to look at it?"

"I see what you're driving at, Dan, but I hope we're not turning ourselves into computers," Maury said.

"We're not. The point is that computers are developing some very human characteristics," I said. "Computer scientists working with neurobiologists have modeled the computer's brain based on the human equivalent. And they call it a database. What we do when we learn and accumulate experience is similar to building a database. Tanya was born in 1960. If I asked her to use a computer built in 1960, she'd laugh at me. But we don't laugh when Brenda, or you Lonnie, or you Maury, try to use a database that is even older."

"What about enduring human values, what about experience?" Maury asked. "They're not worth anything?"

"Worth their weight in gold. Enduring values endure because our experience has taught us their importance. That's where computers and human beings part company. And that's another reason why we are unlikely to be fully replaced by machines. Your ability to acquire experience and learn from it and develop creative solutions, Maury, equips you to run circles around a computer. Whenever the subject of humans versus machines comes up, I like to tell people about endoscopic surgery, one of our new tool cards, and how it is changing hospital operating rooms. Brenda, you probably know something about this technology by the thumbnail sketch I provided the other night."

"A little bit," Brenda said, leaning back in her chair. "The surgeon makes a tiny incision, less than an inch long, and then inserts a long, very thin hollow tube carrying fiber-optic filaments into the body of the patient. The filaments convey a bright light and connect, at the end of the tube, to a minute TV camera that transmits pictures back to a video monitor that the surgeon watches as he operates. The actual operation will be performed with specialized manipulating instruments that are also inserted through the tiny incision." Brenda frowned, and added, "Or, as she operates," to correct her inadvertent gender stereotyping.

"Not a bad summary," I said. "The endoscope minimizes blood loss and trauma by allowing the doctor to extend his or her perceptions inside the body without having to cut a large opening in the traditional way. The incision can often be closed by a Band-Aid rather than a suture. There's little scarring, and the healing process takes much less time."

"A great way to get health care costs down, lower insurance premiums, and lower the amount of pain and discomfort for the patient all at the same time," Maury added. "But what's this have to do with enduring human values and experience?"

"We'll talk a little more about the endoscope later, but the reason I'm mentioning it now is that this technology really points up the absolute necessity for older, experienced surgeons to update their mental databases," I said. "A young surgeon, one who's just out of medical school, is going to immediately take to endoscopic surgery. It's state-of-the-art. But do you, Maury, want that young man or woman doing the first operation of a career on you?"

"Thanks, but no thanks."

"I feel the same way. I don't want to be wheeled into the operating room and have the surgeon say, 'Mr. Burrus, this is a first. I've never done an endoscopic gallbladder removal before.' I'd much prefer the experienced surgeon who's been in that operating room hundreds of times and was open-minded enough to learn how to use new tools like the endoscopic technology."

There was unanimous agreement with that observation. Samantha extended it to cover airline pilots as well.

Maury nodded. "I felt a lot safer when the majority of our commercial airline pilots were veterans. Those guys knew how to get 'em up there and get 'em back down in one piece. Someone who's only done commuter hops can't possibly match that."

"Those pilots started their careers on single-engine prop jobs and they updated their mental databases until they were flying 747s—my point exactly," I said. "Re-become an expert."

Job Hunting

"Where do you start, Dan?" Samantha asked. "There's so much to learn."

"Start with what interests you. People ask me all the time to recommend a hot career in high technology. The other day a young man said to me, 'How about advanced robotics? That's got to be a terrific career in the future.' And I told him that he was right; it is hot and will continue to be that way. But not if it's something you're going to pursue just to pick up a paycheck. Search for your passion by exploring many options, and a career will find you. I've met many professionals over the years who were in the wrong field. They were obviously looking for a hot job that, because it wasn't a passion, turned into a block of ice.

"Samantha, let's have Tanya answer your question: Where do you start re-becoming an expert?" I asked.

Tanya nodded slowly. "I know what my passion is," she said. "Farming. So that's easy enough. But I need to learn more about biotechnology. As a farmer, I've got to find a way to cut my overhead and—or—raise my profit margin; I don't need to be a business genius to figure that out." She stopped and bit her lower lip. "For starters, forget about the more dramatic applications—minicows and drug factories—and look at things like changes in the care and feeding of livestock. Can I cut my feed bills? Can I avoid handing over 20 percent of my revenue to the vet? Can I eliminate the miscarriages that deprived me of the half-dozen or so calves I was counting on last spring?" She stopped and looked down at the "re-become an expert" card, which was lying faceup on the table. "I don't know, but I better find out."

"And when you do find out," Glenn suggested, "put the information on the E-mail system that you're going to set up for the Farm Bureau."

"I was thinking the same thing myself," Tanya said. "We could use E-mail or voice-mail to alert farmers to the latest developments in biotechnology."

"I like what you're doing, Tanya; go ahead and creatively apply technology. But watch out for using the wrong technology to implement the right idea. You'll get into trouble that way," I cautioned. "E-mail is great if the users are disciplined. The messages have to be short and to the point. The same goes for voice-mail. On the receiving end, the mailbox has to be checked regularly. There's so much happening on the biotechnology front, your new E-mail system would be quickly overloaded."

"Presumably, she is going to be using a computer, a modem, and a telephone for E-mail," Brenda said. "Couldn't she also set up an electronic bulletin board?"

Tanya practically jumped out of her chair. "Great idea, Brenda! Farmers from all over the state—all over the country, for that matter—could dial a number and get on-line to exchange information and post the news of the latest biotechnology breakthroughs. I could set that up and run it from the farm. Some of those electronic bulletin boards charge up to twenty-five or thirty dollars a minute for access time."

"Don't limit it to biotechnology," Doug advised. "Have niches or specialties to cover all aspects of farming. Those folks could get on the cellular phone while they're out plowing and . . ." He stopped, perplexed.

"Problem?" I asked.

"Two of them," he said. "One, most farmers don't have computers, I bet. And two, you can't drive a tractor and operate a computer keyboard."

"Most farmers don't have computers?" I asked in mock amazement and laughed. "Most people in general don't have computers. In 1992, Apple estimated that out of the nation's 66 million families, only 7 million owned computers." I turned to Maury. "Do you have a PC at home?"

He looked very indignant. "Of course I do."

"Sorry if I offended you, but a recent survey of 457 top chief executives found that only one in five, or a little over 21 percent, said they used PCs at home *or* the office."

"Fire 'em," Maury growled.

"Fire 'em up is even a better idea," I said. "If Tanya is going to start an electronic bulletin board for farmers, she's got four things to do first. Otherwise, she will have rough going. I've seen it happen. But first of all, let's look at this." I went to the flip chart and turned a page.

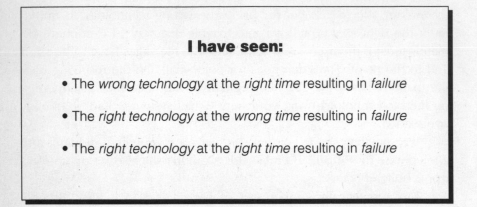

I have seen:

- The *wrong technology* at the *right time* resulting in *failure*

- The *right technology* at the *wrong time* resulting in *failure*

- The *right technology* at the *right time* resulting in *failure*

I pointed to the last line and asked, "How can the right technology at the right time result in failure?" I waited a moment, but no one volunteered an answer. I turned another page of the flip chart:

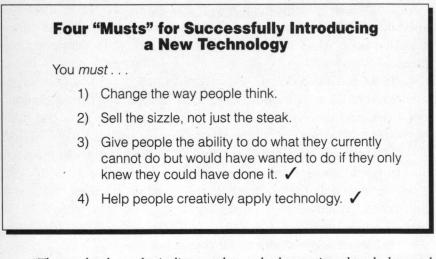

Four "Musts" for Successfully Introducing a New Technology

You *must* . . .

1) Change the way people think.

2) Sell the sizzle, not just the steak.

3) Give people the ability to do what they currently cannot do but would have wanted to do if they only knew they could have done it. ✓

4) Help people creatively apply technology. ✓

"Those check marks indicate rule cards that we've already learned to play with. Now, in Tanya's case, number three really comes into play immediately. She's got to give farmers the ability to do something they currently cannot do but would have wanted to do if they knew they could do it." I paused to underline the rule. "Many people don't have computers because computers don't do anything they want to do—or so they think. But if they start hearing about Tanya's great new E-mail system and the useful information it's providing, those farmers might buy a computer and plug in. Thus, E-mail amounts to an interim technology to move us on to the electronic bulletin board, which, in turn, eventually serves to get us to advanced expert systems."

Tanya grabbed the idea and ran with it. "Like the farmer with those cattle embryos, I could go to a university and work with them to set up an advanced expert system on biotechnology for dairy farmers."

"At that point, Tanya is not only *not* in the farming business—pardon my double negative—she's in the biology and software business," Lonnie observed.

"You've got it. I'll bet she can even figure out a way around Doug's problem with driving a tractor and operating a computer keyboard," I said.

"She—" Glenn started to interrupt, but I cut him off.

"Let Tanya answer." It didn't take her long.

"Recycle the biotechnology information from the bulletin board and put it on an audiocassette. They can play it on the tractor's stereo system."

"Stereo system. You're kidding."

"Believe me, Maury," Tanya shot back. "Tractor stereo systems are to American farmers what the hoe was to farmers in the last century. If it's not built into the cab, those folks use a Walkman. They could listen to the tape as they plowed; it would be like a newsletter. They'd go home and call up an electronic version on their computers and pull out whatever they wanted for their files.

"Tanya is now in the publishing business in addition to biology and software," I said and dealt her another new tool card.

∞

**Multi-
Media
Computers**

∞

"Run that one through your imagination," I urged her. Tanya doodled on her yellow legal pad. She was having trouble. "One way to stimulate your creativity is to ask 'What can I do with the new technology to re-invent an old product?'" I turned to the rest of the group. "Any ideas?"

Doug finally spoke up. "You said Tanya was in the publishing business. She could use multimedia computers to publish agricultural books and newsletters that integrated audio and video and text." He shrugged. "It seems to me that the real business Tanya is in is teaching. She's going to have to teach herself and other farmers about biotechnology and other twenty-first-century technologies. Use multimedia computers to re-invent the concept of a textbook."

"Do you just plug multimedia computers into the phone lines like E-mail?" Glenn asked.

"It's possible, but the phone company's old copper wire network has difficulty handling that amount of data traffic. But fiber optics are already closer to rural areas than many urban parts of the country because it's easier to do the installation away from the congestion of major cities and sprawling suburbs," I said. "The other route is direct broadcast satellites, which would give Tanya global reach. Many U.S. farmers already have satellite-receiving dishes, and the next generation will be three feet across; later in the decade, they will be the size of a paperback book."

Dipsticks and Drumsticks

I turned back to Glenn. "Didn't mean to cut you off a moment ago when we were discussing computer keyboards on tractors, but I could see that Tanya was ready to roll."

"All I wanted to say was that another way around the computer keyboard and the tractor problem is to use pen-based electronic notepads. They'd have to stop the tractor, otherwise they'd end up in the ditch, but farmers could use pen-based electronic notepads anywhere; hooked up to the phone company's network, similar to a cellular phone with a modem, a farmer could access Tanya's advanced expert system and troubleshoot anything from 'What should I do about the oil leak in my pickup?' to 'My chickens look a little sickly.'"

I pointed to my four "musts." "What's happening here? We're

changing the way farmers think—they're in the biology business.

"We're selling the sizzle—a steak is just a piece of dead meat. A computer is a bunch of plastic and glass and wires, but look what it does!

"We're giving farmers the ability to do something they cannot do—fix the pickup truck without calling a mechanic and avoid diseases in the henhouse while plowing their fields." I paused and then asked, "What's the rule that is being used here?"

Lonnie and Samantha spoke at the same time: "Leverage time with technology."

"Right. And we're helping them creatively apply technology—another rule. The rules and the four 'musts' make technology happen.

"Those sickly chickens that Glenn mentioned," I added. "The U.S. Department of Agriculture is working on sensors that can monitor the physical condition of livestock. The farmer can be on a tractor out in the north forty, and the sensors could inform an electronic notepad that the chickens are running a temperature or are getting dehydrated. The warning could also be flashed directly to the vet, who in response either visits the farm or sends instructions directly to the electronic notepad."

"I wonder if I could get the vets to invest in setting up the system? It would certainly be a boon for them," Tanya inquired.

"Never mind the vets," Samantha said, "I'll invest in it."

Unfit for Humans

I gave Tanya her choice of cards, and she selected this one:

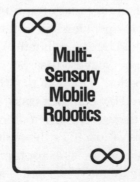

"Let me guess why you picked up multisensory mobile robotics," I inquired. "You'd like to have a robot muck out the barns."

"How'd you guess?" Tanya replied with a chuckle. "Of course, it will never happen."

"Never?" I asked, skeptically. "If a 'smart harvester' can be developed to differentiate between ripe or green tomatoes and corn, or a 'smart sprayer' that knows the difference between weeds and crop plants, mucking out the barn is a cinch. Smart machines—remember smart bombs?—using computers and sensors are going to be widely available in just a few years to perform dozens of traditional agricultural tasks."

"Just the barn would be fine with me," Tanya said.

"Hey, smart machines belong in smart—and smelly—barns. A smart machine that can move is another term for *robot,* and robots are important new tools of the twenty-first century for just that reason. They'll be doing jobs that are unfit for human beings. Not many people want to spend hours shoveling cow manure. The robot doesn't mind at all. Any dirty, demeaning, repetitive tasks are perfect for the robot."

"Dan, I've been hearing about robots since I was a little kid," Glenn said. "Talk about *Waiting for Godot.* What's it going to be, another hundred years?"

"Early robots were primitive and had few practical uses," I said. "Development was limited by what computers could and couldn't do. Just getting the machine to move across the room and beep required enormous amounts of memory and power. I was talking about UNIVAC I earlier. UNIVAC I's digital grandfather was called M1; it was built at Harvard in 1949. M1 could multiply a twenty-one-digit number in three seconds. Of course, it took 500 miles of wire and 750,000 components to do it. That baby wasn't going anywhere! But today's supercomputers are ten million times faster at a fraction of the size. A 1992 chip performs electronic operations as fast as 4-billionths of a second. By the year 2000 it should be 200-trillionths of a second."

"Those figures alone suggest that we haven't seen anything yet when it comes to robotics," Doug said. "That processing speed will allow robots to do incredibly complex tasks, I'd say."

"Like reprogram themselves when things go wrong, or have a three-dimensional sense of touch, or the ability to hear sounds, to analyze what they mean, and to take appropriate action," I explained. "Mobility is improving, and with sophisticated sensors and input devices—a bar-code reader, for example—a whole new era of robots is about to begin."

"That's amazing," Brenda said. "Just add up the number of jobs you could give the robot to do around the house or the office or the farm."

"Or the factory," Maury said. "We installed some robots, but they weren't the be-all-and-end-all the consultants made them out to be. Maybe we were expecting too much of a primitive system."

"Either that, or you were trying to roboticize and automate an already outmoded process and product and blaming it on the robots," I suggested. Surprisingly enough, Maury didn't even blink at that implied criticism.

I was about to end the session when Lonnie complained that I had not kept my promise to talk about combining animal and vegetable genes using recombinant DNA technology.

"I want to know if I'm going to have to start walking my dogwoods or providing kitty litter to the pussy willows," he said.

The pun drew a well-deserved groan from everyone at the table.

I stood up and moved to the head of the table. "Most genetic engineering applications in the field of agriculture will be directed at weed, insect, and virus control." I gestured toward Lonnie. "No Alpo for the dogwoods. Also, the researchers are working to alter how plants respond to growing conditions. Along the way, some old products will be re-invented. Tobacco is perfect for this. The highly profitable but in many cases deadly plant can be turned into a cancer-fighting lifesaver. A company called Biosource Genetics has been working on a process to spray tobacco leaves with a solution containing recombinant DNA molecules packaged in a protein coating. It gets into the plant's system and stimulates the production of the anticancer drug interleukin-2, or interferon. Ironic, isn't it? Tobacco farmers of the near future will be growing a plant that will produce cancer-fighting drugs to save many of the people that smoked a different version of the same plant."

There was an amazed silence. Finally, Tanya spoke up. "I need the name and address of the companies involved."

I adjourned the session, knowing that Tanya Conte was going to hold 'em all right, but she was no longer in the farming business, she was in the biology and information business.

7

This Way Out

When to Fold 'Em

"I've been depressed since last night," Lonnie said as we moved toward the dining room.

We were all ready for a punch line to one of his puns, but he surprised us. "You told Tanya to hold 'em," he continued, looking at me. "That's probably what I should have done at the milling machine company. After twenty years, they offered early retirement, and I folded to start my landscape design firm. I wonder if I did the wrong thing."

Maury responded faster than I did: "Do you like what you're doing now?"

"I love it."

"More than your job at the milling machine operation?"

"Infinitely more."

Maury stood up, removed his dark blue worsted wool suit jacket and hung it on the back of his chair. He sat back down again before replying. "If that's not the key indicator that it's time to fold 'em, you'll never convince me."

The CEO was right, but I knew what was troubling Lonnie. The new cards are so dazzling and powerful they seemed like a cure-all. "We've been using a rule, *Leverage time with technology,* almost since our first game," I said. "If you had been able to play that card in your old job, what would have happened?"

"Ideally?"

"No, realistically. What would have happened to that extra time?" I asked.

"We would have made an outmoded milling machine but made it faster." Lonnie smiled. "As a result, I would have had more time to handle customer complaints and to do the jobs of two other workers who had been laid off."

Brenda was sipping iced tea. She thumped the glass down on the table and said, "And you don't think that it was time to fold 'em?"

"No, it's a valid point, Brenda," Glenn countered. "When do you know it's time to fold?"

I went to the flip chart.

When to Fold 'Em

1) Does your competition have a unique strength you can't match?

2) Is your product or service about to be rendered obsolete by something new that is outside of your area of core competencies?

3) Do you spend more than half your time solving the same types of problems over and over again?

4) Do you work harder and longer to accomplish the same ends?

5) Are you expected to do more with fewer resources?

6) Do you and your organization have a positive vision of the future of the industry you are in?

7) Do you look forward to going to work in the morning?

"These seven questions are a good, quick reality check," I said. "The last two might sound obvious, but I think they are perhaps the most important. If the answer is no to the last two, something is wrong. We've got to get past this notion that work is *not* something to be enjoyed. Trust your instincts. If they're saying 'I hate this job,' it's a very strong indication that it's time to fold."

"Which comes first, Dan, 'I hate this job' or the lack of a positive vision of the future?" Doug asked.

"Good question. Oftentimes, we get ourselves into dead-end situations because there is a lack of a positive vision of the future to start with. We take the first thing that comes along or make the best of a bad deal instead of dreaming and then making those dreams come true with a plan of action built around the new tools and rules of technology. But the opposite may also be true. A bright, ambitious newcomer arrives on the scene. He's got big plans, high expectations and—ten years later—malaise and disillusionment have set in."

"Sounds like me," Doug admitted ruefully.

"Me too," Lonnie said.

"What happens in that case is the once positive futureview became frozen in time. It's ten or twenty years out of date," I said. "There's no change built into the plan or into the futureview. The new tools and rules—the new reality—have all been ignored.

"Look at a positive futureview as the equivalent of the canary in the coal mine. Miners would take canaries into the pits because the birds would keel over in reaction to very small amounts of methane gas. When the canary stopped singing, it was time to get out of there. When your positive view of the future 'stops singing,' it is a warning that it may be time to fold 'em.

"Socrates said, 'The life that is unexamined is not worth living.' Examine your life, your job, your situation with an eye to determining how, under existing circumstances, you can develop a positive view of the future. If you make an honest effort at it and still come up without a positive futureview, it is definitely time to fold 'em."

"When you put it that way, Dan, my decision to fold 'em at the milling machine company was based on a similar analysis. I couldn't see how things were going to get any better—but I didn't know about the new technological tools. Maybe that would have changed the outcome."

"Take any of the tool cards that we've worked with so far and tell me how it would have improved the situation at the milling machine company," I requested.

Lonnie thought it over. "EDI would have been a help. Our suppliers were forever lousing us up by not shipping on time or only filling half an order."

Glenn shook his head. "If your product is obsolete, as that milling machine apparently was, the most elaborate EDI system in the world isn't going to help," he said.

Maury jumped right in. "So what if you know that the supplier is going to have the new fuses ready by the time he promised? Big deal, you don't have to apologize to the customer for not getting him the machine promptly. But the machine he gets right on time belongs in a museum."

"You're both right," Lonnie said. "Maybe we could have figured out a way to update the machine and keep it alive for another ten years."

"How about using an advanced expert system to give the miller a new lease on life," Tanya suggested.

"Let me give you an example of a company that tried to do something like that," I said. "They didn't have advanced expert systems, but RCA, back in the 1940s and 1950s, did it the traditional way. The managers and engineers got together and decided that they weren't going to give up on their cash cow, the vacuum tube, even though their competition was coming on strong with transistors. They poured millions of dollars into developing the next generation of vacuum tube. The ultimate in sophistication. What happened? Advances continued to be made in solid-state technology. Transistors proved to be faster, cheaper, and more reliable. RCA woke up one day and, yes, they were leaders in vacuum-tube technology, but they were nearly out of the running in a technology that would drive the industry for years to come."

"That story hits uncomfortably close to home," Maury said. "We're spending millions upgrading our mainframes, and what you're suggesting is that we're building a better vacuum tube."

"I'm afraid so, although it will take a long time before the market is dead. Dying? Yes. Dead? Not yet. It's a pattern that can be seen throughout history. As new technology comes along to supplant the old, there's a tendency to invest heavily in an effort to salvage an obsolete product or process. Clipper ships arrived just as the steam engine was about to revolutionize merchant shipping. The clippers were beautiful and fast; they gave windjammers a ten- or fifteen-year reprieve. But in very short order steam engines sent those magnificent vessels, and the magnificent investments that had been made in them, into oblivion."

"I guess you can run but you can't hide from the new technology," Brenda remarked.

"And the running part can be fatal," I said. "It squanders resources that could be used to develop new products and to re-invent the company. When you fold 'em, it's not necessary to close up shop and go out of business every time. Take the natural gas industry. Back in the nineteenth century an enormous investment was made in plants and a network of pipes to supply homes and offices in big cities with gas for illumination. Just when they were sitting pretty, Thomas Alva Edison hits town and gaslights are finished. Even so, the gas companies didn't just fold 'em; they re-invented themselves as suppliers of gas for cooking and home heating. If they had tried to fight it out with Edison for the lighting market, it would have been a massacre."

"Brenda used that great old Joe Louis line about how you can run but you can't hide," Maury said. "American companies have been running overseas for the past twenty years or more. In 1992, Smith Corona, the last U.S. typewriter manufacturer, announced it was heading for Mexico. They've probably spent millions of R&D and retooling dollars to add a small amount of memory and a three-line display into their typewriters—a perfect example of trying to keep a relatively obsolete product from dying. Now, they've relocated in Mexico." Maury shook his head. "They don't need a new country, they need a new product."

Reoccurring Problems

"To go back to that chart, Dan, if you wake up in the morning and say, 'I hate my job,' the first thing to do is examine your futureview based on the impact of the new cards," Glenn said. "If it isn't positive, you're in trouble—and the reason it isn't positive is probably because change wasn't built into it. Am I basically summarizing correctly?"

"You are," I agreed, "but keep going. Do a full assessment. Is change built into your job, into the product, the process, and the company's plan? If the answer is no, then those last three questions on the chart come into play.

"When the same problems keep showing up again and again, it means that you're going nowhere—fast. That's another solid piece of evidence that it is time to fold 'em."

"Actually, Dan, doesn't that harken back to the rule *Take your biggest problem and skip it?*" Brenda asked, and then the hospital information manager proceeded to answer her own question. "Fold 'em and find a new product or process or plan. That way you've skipped your biggest problem, which otherwise would never go away."

Doug shook his head and frowned. "That may be easy to do in business or a hospital, Brenda, but at school there's no way to skip our biggest problems. One of the worst problems we have with science education is teaching kids math skills. Science and math are dynamically linked. But it is an incredible chore. Because of our current tenure system, many teachers who teach math or science hate it, and as you could guess, the kids hate it too. But we can't skip it."

"Yes, you can," I said. "Skip the problem by giving the job to a multimedia computer. Remember, you suggested it to Tanya as a way to re-invent the textbook. Use it to re-invent the failed mathematics teachers. Stop beating your heads against the wall. Just like the robot that's going to muck out Tanya's dairy barn, give a machine the dirty work to do. That multimedia computer, with sound and video and text, can give math all the pizzazz of a rock video on MTV."

"Why not just get rid of teachers altogether?" Doug asked without looking at me directly.

"Because computers can't do what teachers do best. It's true that a good multimedia computer program can motivate, thrill, and engage a student. But teachers who have passion for their subject and a personal interest in their students' success have a very powerful teaching gift computers lack—emotions. For example, your love of science is contagious. The only way that love will spread is through human contact. Computers, no matter how advanced, don't love or care for the subject or have an interest in the students they teach. Kids know this and react to it. A computer, no matter how advanced, cannot play the part of mentor and role model.

"Obviously, it's a waste of resources to ask teachers to do a job that's more effectively performed by a machine. Or, to expect a machine to look a child in the eye and say, 'I knew you could do it.' So what happens? The choice seems to come down to encouragement or coercion. And since machines aren't very good at doing either one very well, the human gets saddled with coercing the student, the parents, the workers, or the customers (seeing that encouragement takes so much time and effort).

"In the first place, one of the mistakes we make in schools and businesses and hospitals is assuming that coercion is the answer. It isn't. And then—because coercion fails, as it must—we fall back on good old-fashioned persistence: If at first you don't succeed, keep doing it again and again—only harder."

"Try, try again—but use the new tools," Glenn added. His statement was immediately fielded by Lonnie and lobbed back.

"Try, try again—as long as you're not using cutting-edge technology on behalf of a rusty razor blade of a product or process," he said. "You just slit your own wrists that way."

"General Motors lost a lot of blood with the rusty razor blade approach," I said. "GM put robots in its existing production facilities over a decade ago and created a very expensive problem. It spent $40 billion on modernization. They automated an existing system, for the most part, that wasn't initially designed to be automated with robots. They should have folded the plants and started from scratch."

"That would have been hellishly expensive, Dan," Maury replied.

"It was hellishly expensive *not to*. GM management was great at doing what I call 'present-thinking.' They thought about the bottom line in 1982 instead of the bottom line in 1992. If the entire assembly line method using those plants had been terminated and new fully integrated plants built ten years ago, the agony of plant closings, layoffs, and poor financial returns that GM is going through in the early 1990s would have been lessened considerably. It is easier to be a crisis manager in the short run, but in the long run it's imperative to be an opportunity manager. Automating an obsolete system only means that you will fail faster."

I pointed to the chart. "GM installed robots in plants that were built in the 1940s and 1950s, plants, by the way, that are still capable of producing fine automobiles—1940s and 1950s automobiles. Management was out to solve its biggest problems. Anybody want to take a stab at what those were?" I asked.

Glenn raised his hand. "Aside from the big-car strategy that allowed the Japanese to make such inroads, my guess would be declining productivity and quality and increasing labor costs."

"There are some others, but those three will do," I said. "Did the problems go away once robots were introduced? No. If anything, they got worse."

"Time to fold 'em," Tanya said.

"As anyone would know who bought a 1971 Chevy Vega," Maury said. "I got one for my wife—what a mistake that was!"

"The Vega was intended to go head-to-head with the Volkswagen Beetle, just as the 'Bug' was about to give way to the likes of Toyota, Datsun, and Honda," I said. "GM challenged the competition all right—yesterday's competition. When the Vega didn't perform up to expectations, the company went into its Lordstown, Ohio, plant where the car was being produced, laid off 700 workers, and installed high-speed welding robots."

"I remember that one," Lonnie commented. "They had a long strike on their hands."

"Actually, it only lasted three weeks, but it must have seemed like three years to the company. The speed of the assembly line was cranked up to suit the robots, only there were still humans on that line, and they couldn't keep pace."

"Is it any wonder? There were 700 fewer of them to begin with," Samantha observed.

"The unions walked off the job in protest, and before the dispute was settled, the Vega's reputation was mud and the public was convinced by the news media that modernization in general, and robots in particular, were going to destroy the American way of life."

Maury got off one of the better lines of the evening: "As I recall, the Lordstown strike did for robots what the Hindenburg did for blimps."

"It may not have been quite that disastrous," I said, "but *robot* is still a four-letter word to many workers."

"So is *V-E-G-A,*" Glenn said.

Speed Bumps

"Looking at your chart, Dan, and thinking about Lordstown suggests to me that those last two points were big factors," Lonnie said. "The workers were working harder and faster to accomplish what had been done routinely only a few years before. And that final one—do you look forward to going to work in the morning?—is clearly involved: Lay off 700 workers, and you are asking the survivors to work with fewer resources. We didn't have robots, but the vicious circle of layoffs and more work with fewer people is exactly what happened at my old company, and exactly the reason why I bailed out to start my landscaping design firm."

"I'm sure GM figured the robots were an additional resource," Maury said. "And an expensive one at that."

"But 700 people . . . " Lonnie turned to me to ask, "How many robots?"

"Twenty-nine."

Doug whistled in amazement. Lonnie continued: "As I was saying . . . fewer resources."

"Particularly since the robots had a habit of breaking down, which meant that the human workers had to be there to perform maintenance and pick up the slack. Plus, they weren't in the best of moods, given the massive layoff that had occurred. Workers didn't have much of a personal interest in seeing the modernization program succeed.

"Put GM aside for a moment and return to the milling machine company or the mainframe manufacturer or the hospital," I said to Lonnie, before turning to Doug and Samantha. "Don't feel left out," I told them. "I could have also added the insurance industry and education. The Lordstown experience is particularly vivid, given the circumstances. We can picture those workers rushing madly to keep up with the robots. But the same thing is happening in almost every company and organization. Imagine that we've filmed or videotaped the typical office and then played it back at high speed. People would be whipping between the desks and the filing cabinets, grabbing the phones, hanging up, and rushing to the copier. Bedlam. Welcome to the office environment of the 1990s!

"Almost all of us feel stressed out and overwhelmed. And what is that telling us?"

"Time to fold 'em," the group said in unison.

"It's a dead giveaway. When I start to feel the stress mounting, I know right away that I've got to re-evaluate the situation and begin updating," I said. "My tools and processes are simply obsolete and no longer functioning properly."

Lonnie patted his chest. "For the last two years I thought I was on the verge of a heart attack," he said. "There was no way I could get all my work done in eight hours, and I'm very organized and methodical."

"So am I," Brenda said, "and I'm in the same boat. Just to clean my desk off at the end of the day, I'll spend an extra hour or ninety minutes."

"Clean desk? What's that?" Samantha asked. "I'll never win the insurance industry's coveted Good Office-Keeping Seal of Approval."

In response, Lonnie got out of his chair and stood behind Brenda. He put his hands on her shoulders. "Are we saying, then, that this woman should fold 'em and get out of the hospital?" he asked and moved behind Samantha. "Are we saying that this woman, as young as she is, should fold 'em and get out of the insurance business? In both cases, there's evidence that it's time to fold 'em."

I deliberately waited before answering. I wanted to see if someone else would tackle the question. Tanya came through like a champ. "A positive view of the future, Lonnie—you're forgetting that," she said. "I had the same blind spot. Farming looked like a lost cause until I did what Dan suggested and started to look at the new big picture and use these new rule and tool cards. You admitted a little while ago that you couldn't see how the new tools or rules could have been used at the milling machine company to pull the operation back from the brink."

"In my case, Lonnie, I still like going to work, no matter how long the hours," Brenda said. "I think the problems are surmountable—they've got to be."

Samantha followed up by saying that she could come up with a half-dozen positive views of the insurance industry's future using the new rule and tool cards.

Lonnie shrugged, went back to his place at the table, and sat down.

Multiple Choice

Retrain, Relocate—or Retire

I used my pointer to emphasize the new page of the flip chart. "I think it was social futurist Marvin Cetron who spelled out the options like this: 'Retrain, relocate—or retire.' There's a lot of truth in it.

"Lonnie is living proof. He could have taken his pension and retired. Instead, he retrained and relocated out of the milling machine operation. Without analyzing it closely, he realized that there was more of a future as an entrepreneur, as a landscape designer, than staying in the same old rut.

"The entrepreneurial spirit is what makes America unique and, ultimately, unbeatable. The reason is that it is based on hope, hard work, and a positive view of the future. And that's 'how the West was won' after all. Even so, like farmers, there's a streak of the gambler in pioneers and entrepreneurs. They roll the dice."

"Amen to that!" Lonnie said. "Did I ever roll the dice!"

"Now, throw the dice away and pick up the new cards."

"I can't," Lonnie said.

"Why not?"

"Because I've been paying attention. You think that just because I make a bunch of bad jokes that I'm out to lunch," he said. "I saw what you did with Maury and Glenn and Tanya. I'm not going to pick up those cards until I know what I want to do with them."

"Fine by me," I said, well pleased with his attitude. "What do you want to do with them?"

Lonnie shrugged. "Cut grass, plant flowers, trim shrubs."

"You jerk!" Samantha exclaimed.

"Just kidding."

"Well, in that case, you're definitely going to need a multimedia computer," Maury said, flipping him the card.

"No, not that one," Glenn said, joining in the teasing. "Diamond thin-films . . . that's what Lonnie should be playing."

I called the group to order. "Okay. What do you *really* want to do with the cards, Lonnie?"

"Darned if I know," he admitted. "I sounded flippant when I said cut grass, plant flowers, and trim shrubs, but that's pretty close to the truth. Actually, do that, which isn't much fun, and make a profit so that I could spend the rest of my time doing what I like to do—designing. At the moment, I'm subsidizing the operation out of my savings and pension."

Lonnie went on to explain that the competition in the landscaping design business was so intense that he was finding it difficult to break in, even on the ground floor, which featured cutting grass, planting flowers, and trimming shrubs. "The name of the game is getting your foot in the door, but I keep having to cut my prices to win the jobs, and then I run at a loss or a near loss," he said.

"Is that your biggest problem?" Maury asked. Lonnie nodded. "Then skip it," he advised.

"Easy for you to say . . . but all right, I'll skip it."

"Maury's right. Skip it so that you can look twenty years down the road and determine where you want to be," I said. "In that way, you'll bypass this obstacle."

"I'd like to be the oldest—I'll be pushing seventy—and the richest landscape designer in southeastern Connecticut. *Numero uno*." He paused. "No, make that number one in Connecticut."

"First Connecticut and then the world," Doug said.

"No, first Connecticut, then the Hamptons—and then the world."

"Good start," I said. "Now, how are you going to get there using the new cards?"

Lonnie picked up the multimedia computer card that Maury had tossed his way. Shook his head and put it down. He did the same with diamond thin-films. "I need technology that will cut grass, plant flowers and trim shrubs cheaper and faster than I'm doing now. How about multisensory mobile robots?"

"Lordstown here we come," Samantha said.

The comment made Lonnie defensive. "I've tried working my guys faster and harder, paying only the minimum wage. I got sloppy work, high absenteeism, and unhappy customers. Robots won't get me in trouble with the U.S. Immigration and Naturalization Service."

I turned to Maury. "Deal him the multisensory mobile robotics card." He looked through the deck and placed it on the table with a snap. Lonnie went to reach for it. "Now, take every chip that you have and put them in the pot," I said.

"Huh?"

"Yes, the whole pile, and you had better borrow some from Brenda. If you're going to play with high-end technology, it requires a substantial investment."

"Well, I can't do that. I'm already losing money."

"A good thing," Maury said. "Why invest in technology that allows you to do something that you don't want to do in the first place? You said you wanted to design."

"But mowing lawns is entry-level stuff," Lonnie explained.

"With an exit-level price tag if you invest in a lot of expensive technology," Samantha said.

"How else am I going to reduce my costs so that the job shows a profit?" Lonnie asked, exasperated. "I thought that's what the new technology was going to do."

"Is that your goal?" I asked.

"Yes."

"I thought your goal was to be the number one landscape designer in Connecticut and the Hamptons and to go international?"

"Yes."

"Which is it? Yes—cutting costs? Yes—number one?"

"I'm only looking to cut costs so that I can keep my prices competitive, stay in business, and eventually become number one. There's a long-term and a short-term goal."

"But your long-term goal is being jeopardized by your short-term goal."

"So I need a *new* short-term goal then?"

"Don't change the goal necessarily—change the way you think about the goal. The goal isn't cutting costs. The real goal is to make a profit. Cutting costs may seem like the way to get to that goal, but there are better ways," I insisted. "Currently, cutting costs is actually cutting your profit margin. You said that by pushing your crews harder and faster, the work was sloppy and the customers unhappy. The strategy isn't working. Ask yourself what else *will* work using the new tools and rules of technology."

Lonnie looked around the table for help. "Can I give him a hint?" Doug asked. I nodded. "It's like taking a multiple-choice exam. If you don't know the answer, eliminate the obviously incorrect options, narrow the choices to two, and then guess. You've got a fifty-fifty chance."

"Now you tell me. I always flunked multiple-choice tests." Lonnie hesitated. "The only alternatives to cutting costs are to keep them where they are or to raise them. Keeping them where they are is slowly bankrupting me—so that gets ruled out. I've tried cutting them; I guess the choice is raising my costs with the object of raising my profits."

"Bravo," I said. "You've just liberated yourself from the most destructive form of competition—price."

Opposites Attract

"The purpose of these card games," I said, "is not rest and recreation, although we're all having a good time. The purpose is to re-invent the way we compete." I got out of my chair and stood in front of the flip chart with my back to the group. I tore off the top sheet and crumpled it up. I turned around slowly. "I'm going to commit a subversive act and I'm going to do it in writing." I turned back to the chart and went to work with my felt-tipped marker.

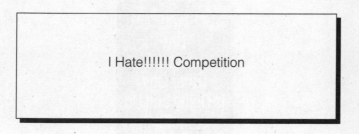

I Hate!!!!!! Competition

"It seems un-American. The United States Constitution requires competition, doesn't it? The answer is no. It isn't mentioned. Maybe the Founding Fathers knew that competition was wasteful, inefficient, and destructive. Many of them were wealthy men who had gone beyond the competition in the colonial era to amass great land holdings and capital."

"But what about the so-called invisible hand of capitalism?" Glenn asked. "Competition is supposed to regulate and energize the marketplace."

"I'm not arguing with the economic theorists. This is pure, practical business that I'm talking about. Lonnie, if I could knock your landscape design competition out of the box, what would you say?"

"I'd say 'Thank you very much.'"

"I thought so." I turned to Maury. "What would you say if I whispered the magic word and the competition disappeared?"

"I'd say 'No thanks,' the competition keeps us fired up and on our toes."

"And I'd say to that—*Bull!* "

The rejoinder drew a round of applause, which Maury enthusiastically joined after a moment or two of looking surprised.

"Let me amend that," he offered. "The competition keeps us fired up, on our toes, and living in fear of the quarterly profits report."

"That's a little better," I said. "If competition is used as a shorthand way of talking about fairness and hard work, I don't have any problems with it. But if competition means a pack of mangy dogs fighting over the same bone—forget it!"

I leaned over the table to reach for the pack of rule cards. "A tremendous professional speaker and friend, Joel Weldon, had a great way of describing a key rule for moving beyond competition." I held up the card.

I gestured toward Lonnie. "What you're doing is just the opposite. The other landscape design operations are cutting grass, planting flowers, and trimming shrubs—so that's what you do. The others cut their costs to cut deals—so that's what you do. Now, ask yourself what don't they do—and go do it."

Lonnie chuckled and picked up the pack of new tool cards. "I almost forgot this little baby. What my competition doesn't do is play with this card," he said, tossing it out.

"While my competition is busy cutting grass, I think I'll cut the noise out of my customers' gardens," he said.

"First, you have to come to my hospital and pacify the neighbors who have to live with the noise of the sirens—remember?" Brenda asked.

"That's why I thought of antinoise. We were discussing it when Dan first introduced the cards. I immediately had the crazy idea that I could sort of specialize in areas that abut hospitals with the same kind of problems."

"What's crazy about that?" Glenn wanted to know. "My cousin lives across the street from a rescue squad. He'd go for antinoise in a minute."

"Particularly a minute after 2:00 A.M.," Doug added.

"Dan, let me ask a question about antinoise," Maury requested. "Is it like what they call 'white noise,' a neutral, soothing sound that drowns out a barking dog or the neighbor's air conditioner?"

"No, it's a mirror image of the original sound wave, meaning that its contours are exactly the same but reversed—up is down, down is up—and it cancels the other out."

"So all sound is blotted out?" Tanya inquired.

"Just the sound that we choose to cancel. The device takes a sample of the lawn mower or the airplane engine, duplicates it, creates the opposite sound wave, and then transmits it at the same volume whenever the neighbor is mowing the lawn or the jet flies over. If someone honks their horn or knocks on the door, those sounds are unaffected because they are at a different pitch."

"Perfect. I've got a pasture right by a busy road, and the truck traffic disturbs the animals. I was thinking of selling it, but antinoise might be able to eliminate the problem; plus, I could still call the cows for milking by ringing a bell my father found years ago in an abandoned church."

Lonnie slapped his palm on the tabletop. "Roads. Route 95 runs right through Connecticut, top to bottom along the coast; it's some of the most valuable real estate in the world. The noise is almost a constant drone in many areas. I could just work both sides of that highway for the next twenty years—backyards, office parks, golf courses, restaurants, and motels. Just call me the Connecticut Hush Puppy."

"Call me Mr. Muffler, and say good-bye to Midas," Doug said. "Could antinoise be used on a car or truck?"

"That would be one of the easier applications," I said. "It's a single sound source and it's relatively constant. Even better, unlike a traditional muffler, which is a drag on the engine's power, antinoise doesn't get in the way of performance and fuel efficiency. The extra mileage might even pay for the device. There are companies already working on this type of application, but the field is wide open because the technology is so new."

As I was finishing my answer to Doug's question, Maury made a comment to Glenn, who was sitting on his right. Glenn was nodding vigorously.

"Maury obviously hit a home run, Glenn," I said. "Let us in on it."

"Go ahead, you tell him," Glenn said to Maury.

"I said that we should see a big breakthrough in muffler technology in the next few years as the muffler manufacturers try to head off the antinoise people from stealing their cash cow."

"Anybody want to buy a muffler franchise?" I asked. "Cheap."

"The muffler people should play this card," Doug said as he rummaged through the deck. "It's one we've already used."

"Render theirs obsolete and then run off with Lonnie's," I agreed.

"No way. I'm already working on the next generation. It will be antinoise built into pink plastic flamingos. Midas will never top that."

"You're forgetting biotechnology," Tanya said. "A snip of the molecular scissors and the rosebushes will cancel out the sound of passing fire trucks."

"I wouldn't invest in Jackson and Perkins, the mail-order rose supplier, based on that prediction," I said.

"Tanya's joking," Lonnie said, "but I've been thinking about biotechnology since the last time we were discussing it. I should check with the genetics people and see whether it would be feasible to develop a variety of grass that doesn't need cutting after its initial growth spurt in the spring." Lonnie used his hands when he was speaking, and he almost knocked over his coffee cup. "Talk about stealing a cash cow! And if you could do the same with shrubbery and trees, it would be a gold mine for whoever owned the patent. There'd be a licensing fee for every box of grass seed or sapling that was sold."

"Presumably, you could also eliminate all the excess fertilizer and water that goes into lawns at the same time," Brenda said. "And I'd get a chance to see my lawn-slave husband on the weekends."

"Not bad," I said. "You've picked off the landscaping industry's cash cow, one of the chemical industry's cash cows, and you've solved your customers' predictable future problem today by recognizing that water shortages and pollution are going to be increasing concerns."

"And you're leveraging my husband's time with technology," Brenda added.

Show Biz

"How about designing, Lonnie? It's your passion, but you haven't addressed it yet in terms of the new tools and rules," I said.

"Maury was kidding me about the multimedia card earlier tonight—maybe that's one I could use."

"In what way?" I asked.

Lonnie pondered the card. "Help. I'm blanking out," he said. I looked around the table and could tell that Samantha and Glenn weren't blanking out, they were ready with ideas, but I wanted Lonnie to keep trying. We waited in silence. "Got it," he said. "I'm now solving one of my problems—lack of a track record. I haven't been around for twenty years to build a huge portfolio of work. I could have prospective customers come into my office and get a look at my portfolio on multimedia computer—sound, video, drawings, text. Even though there's not a lot of quantity, the quality would come roaring through."

"That's similar to what I had in mind," Samantha said. "Lonnie could use multimedia as a catalog of his services, because that's the way *I'm* going to use multimedia to introduce my customers to insurance products that they would buy if they only knew what was available. Brochures and fact sheets just don't do the job."

"My idea was to actually do the design presentation with multimedia," Glenn said. "Do a sound-and-light show instead of unrolling a blueprint. What customer could say no if they saw an aerial shot of the garden that got closer and closer until they were at ground level. There would be a video of the flowerbeds and trees that had been selected; a text explaining the maintenance program; audio of the songbirds that are attracted to those plantings."

"All right, keep going and combine multimedia with this card," I urged Lonnie.

"What am I going to do with that?" he said in despair.

"Play the cards in combination," I said. "There are three of them on the table."

"Let me figure out what to do with this one first before I start getting fancy. Advanced simulations is like virtual reality?"

"Advanced simulations can be in 2-D or 3-D and displayed on a regular computer monitor or TV set, or they can be in 3-D using a virtual reality hardware and software system," I replied.

"Hah, this is another one we talked about when you introduced the cards. You said that with virtual reality a customer could walk down the path and pick the roses before we even built the garden—right?"

"With a computer for running the software, data gloves to allow you to manipulate simulated objects, and a special helmet with built-in flat-panel displays for observing the simulated world, you could take them to the grape harvest in Bordeaux, France, or go for a visit to the famous White Garden at Sissinghurst in England," I said.

"Wouldn't it be cheaper to fly to London on the Concorde and rent a Rolls for the drive to Sissinghurst," Maury asked, "than to bankroll a virtual reality setup?"

"At the moment, it will be difficult for Lonnie to afford virtual reality. But as the technology evolves, it's going to be within his reach. What he—and the rest of you—should do is track these high-end, cutting-edge technologies. By following them closely, you will be able to know when they are ready to go. If you just assume that this technology is out of the question because of the high current cost, it will wind up in the hands of your competitors.

"Take a look at this next chart."

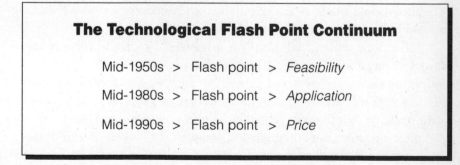

The Technological Flash Point Continuum

Mid-1950s > Flash point > *Feasibility*

Mid-1980s > Flash point > *Application*

Mid-1990s > Flash point > *Price*

"What it is telling us is that there are points in time when technology clears major hurdles. I call those flash points. By the mid-1950s, many of our twenty core technologies were advancing to the extent that it became feasible—expensive, difficult, but feasible—to start developing realistic and practical applications. That feasibility was the result, in large part, of advances in computer technology. Those who were able to work with the core technologies back then were members of a very exclusive club: the biggest corporations and research organizations. It took big bucks to join. Come the mid-1980s, those applications were available to the high-end user. Still, for the most part, a big-bucks proposition, but not *as* big as it had been in the 1950s. Again, advances in computer power and speed were igniting the flash points.

"And what's been happening since the mid-1980s? More speed and more computer power—a lot more. The cost of computing has dropped a thousandfold in two decades. And there's more to come: AT&T has a megabyte chip that will reduce memory storage costs by 75 percent. By the mid-1990s, price, which determines the availability and accessibility of technology, will fall to levels that put the twenty-four cards within reach of small businesses and individuals. The most significant flash point is therefore the price flash point because it is reached only when there's a large number of different applications and the price is no longer an entry barrier. The price flash point will create a revolution if the new technology satisfies a major social need.

"The exact pace between flash points varies. CDs—music and data on CD-ROM—began their application flash point in 1982 when 800,000 units were sold; their price flash point occurred in the early 1990s when most potential users were no longer deterred by the cost. In 1991 333 million units were sold, and in 1992, for the first time ever, reference books such as encyclopedias and dictionaries on CD-ROM outsold the paper-based book versions.

"The same pattern will hold for advanced simulation technology. Virtual reality hasn't triggered the price flash point yet. It still needs enormous computer memory, power, and expensive accessories. But 2-D simulation is almost there. Not as cheap as a typical business software package, but the cost is justifiable to an entrepreneur like Lonnie. He can show customers their garden project on a screen of a multimedia computer in two dimensions.

"How will it appear looking south? There it is. Looking north? You've got it. Every aspect of the garden will be stored as video on the various tracks of a CD that is plugged into his computer; if the customer wants to stroll past the clematis vines that are around the corner from the lilacs, what he or she will see on the computer is the path wending along with beautiful purple lilacs on both sides, and ten yards or so later turning to the right to show the pink clematis flowers. Alternatively, to inspect the tulips, the path would turn left."

"No more drawings that make oak trees look like splotches?" Lonnie asked.

"The oak trees can be made to shed their leaves to show what it will look like in the winter. Or to grow in five-year intervals to give the customer an idea of how the garden will mature," I said. "See the power you get by combining the cards? Multimedia and advanced simulations work well together."

Lonnie leaned back in his chair and said, "Yes, and I could demonstrate antinoise as well by letting the user hear how noisy the garden was because of the road and then flipping a switch and neutralizing out the racket."

Picture Perfect

"Give me another card. This is fun," Lonnie said.

"Which one would you like?" I asked.

"Any of them."

That was the answer I wanted to hear. Lonnie had hit his own personal flash point. He knew that he could play with any of the new cards if he just gave it a try.

"Here's one that Glenn drew," I said.

"I'm going to sound flaky now, but what the heck," Lonnie said. "You told us to figure out what it is that we actually need to do. I guess that means we should question all of our unquestioned assumptions." Lonnie held up the new tool card. "I'm a landscape designer. What exactly is a landscape designer? Can anybody tell me?"

Doug took a stab at it. "A landscape designer is someone who arranges or rearranges elements found in the natural environment," he said, but he immediately shook his head in dissatisfaction with the definition.

"It's tough to nail down. But I think what I do is similar to being a sculptor. Better yet, I'm a graphic artist using the face of the earth as my canvas. So, if I'm an artist, selling my art, which is—I'm sure you'll agree—relatively stationary, why don't I use these flat-panel displays as another form of canvas to provide mobility and reach a wider market?"

"I don't think I'm following you," Maury said.

"What's in the lobby of your headquarters building, Maury?" Lonnie asked.

"The lobby? We've got a big Miro tapestry on one wall opposite the elevators. Is that what you mean?"

"Yep. How would you like a three-story projection of a gorgeous maple tree, leaves rustling in the breeze, sunlight filtering through the branches, the quality of the shadows and light changing with the day and the season, leaves turning slowly red in the autumn?"

"Sounds nice."

"Advanced flat-panel displays," Lonnie said, looking mighty proud of himself.

"Have you been peeking at my notes, Lonnie?" I asked.

"How dare you," Lonnie replied, pretending to be offended. "If Glenn can use flat-panel displays for training and product promotion, I can use them to hang my products on walls all over the world. Why should I just work on somebody's backyard or front lawn? I'm an artist, not a landscaper."

Hard as a Rock

"You're on a roll," I said. "Try this one."

"Thanks, Dan, I was afraid you'd do that. And it's all Glenn's fault for mentioning diamond thin-films in the first place." Lonnie scratched his head. "I'm stumped."

"That's a relief," I said. "I don't want you guys to catch on too fast or the fun will end too soon. Who wants to refresh our recollections about diamond thin-film?"

Brenda raised her hand. "I've got it written in my notes," she said, leafing back through her yellow legal pad. "A process that can deposit layers of diamonds as thin as a single atom onto almost any surface. You could put down a layer of diamonds a millionth of an inch thick."

"Terrific. What's Lonnie going to do with diamond thin-film?"

Maury cleared his throat. "Maybe I can come up with something by looking at it from the perspective of solving the customer's future problems. Landscape design is not just plants; Lonnie installs lighting, garden urns, benches, fountains. A tree will grow for seventy-five or a hundred years. A bench will rust out or rot out long before that. Coat those things in diamond thin-film and they'll last forever."

"Now, why didn't I think of that?" Lonnie said, picking up the diamond thin-film card. "I could coat my antinoise pink plastic flamingos in diamond thin-film."

"Forget the flamingos," Glenn said. "Coat every barbecue grill in the country and you'd be a zillionaire."

"Coat every car in the country and forget about rust and scratches," Samantha suggested. "He'd be a multizillionaire."

Every one of them had an idea for diamond thin-film. They left my house still talking about it, still coming up with new ideas for playing with the new tools of technology.

8

Avoiding Self-deception

Calling a Spade a Spade

I spent the better part of an entire afternoon working on what was supposed to be a list of troubled major U.S. corporations. The plan was to have it ready for the evening's card game. What should have been an hour of work turned into a vast undertaking.

The name of virtually every Fortune 500 company wound up on the list. Some were in worse shape than others, but even the strongest had a common problem: At best, they were playing with a few of the new tools and rules, not the entire new deck.

The story was the same for every industry and sector of the economy. I started thinking about the television networks and wondered if they truly realized how CNN had stolen their cash cow, news programming?

The cable television business, for that matter, was another vivid example: Their transmission system is being rendered obsolete by fiber optics, direct broadcast satellites, and digital microwaves, yet many of the cable companies keep happily stringing up traditional cable. Don't they realize that 80 percent of all households in the

United States are now within fifty miles of a fiber-optic trunk? That's not even accounting for wireless transmissions using microwave technology.

Mile High Cable spent over $77 million and four to five years cabling Denver. A competitor, TVCN, using a microwave, did it in four to five weeks at a cost of $2.85 million.

Far too few corporations are seeing the new big picture in its totality. As I sat at my desk staring at the sheet of paper, I couldn't help recalling a visit I made to the headquarters of Southwestern Bell a few years before. There were video screens all over the building displaying up-to-the-minute stock quotes of Southwestern Bell and the other "Baby Bells." One of their managers had just gotten through assuring me that top management was trying to get their employees focused on the future. But those screens had a different message entirely: Forget about long-term thinking, keep your eyes on today, we may be up a half point by closing time! Subliminally, what the workers were getting as they walked the hallways was confirmation that present-thinking was management's preference.

And if one of the Baby Bells is stuck in a present-thinking rut, and those are companies that are considered to be wizards of high-tech communications, what about more conservative U.S. firms? By the time my seven card sharks minus one (Tanya couldn't make the game) arrived, I had torn up my list and written these words on the flip chart.

The Foundering 500

"Remember when I told you that change isn't what it used to be?" I asked the group. "The Fortune 500 isn't what it used to be. That bastion of American corporate success is looking pretty shaky. Between 1987 and 1992, 143 companies have been replaced on the Fortune 500. That's a casualty rate of nearly 30 percent. Between 1955 and 1980 only 238 corporations dropped off the list. The pace of the attrition rate is running about three times faster."

"Will the last one to leave please shut off the lights," Lonnie said.

"Doom and gloom. All I hear anymore is doom and gloom," Maury complained. "There's got to be some good news!"

"You want good news? Some companies are beginning to use supercomputers in an effective way," I said. "This next chart proves how profitable it can be to *leverage time with technology,* a rule card I introduced you to at one of our first games."

Du Pont's New Material Design

BEFORE	NOW
5 years, $5 million	2 weeks, $25,000

"It used to take five years and $5 million to come up with a new material. Now it's two weeks and $25,000. That's a savings, thanks to the supercomputer, of $4,975,000 per design. The money goes right to the bottom line. The time that's saved makes Du Pont better able to respond to the changing marketplace, and less inclined to stick with obsolete products to justify the expensive research. Take at look at Alcoa's experience."

Alcoa's New Can Design

BEFORE	TODAY
6 months, $100,000	2 days, $2,000

"I'll bet those supercomputers are their best performing assets. The technology probably pays for itself in a few months," Glenn speculated.

"Would you believe that the technology is paying off for Corning to the tune of more than a million dollars a month?" I asked.

Corning's New Glass Materials Development

BEFORE	*TODAY*
20 years	3 months

Saving: $15 Million a Year!

"There are other examples of increases in speed using computers and assorted new technologies. GE used to take three weeks to produce a custom-made industrial circuit. Now it takes three days. AT&T used to need two years to design a new phone, now it needs one. Motorola used to turn out electronic pagers three weeks after the factory got the order. Now it takes two hours."

"See, there is good news," Maury said.

"But some bad news for you, Maury," I added.

"Oh?"

"The $500 million Stealth bomber was designed piece-by-piece on a network of $10,000 desktop workstations, not a mainframe."

"Thanks, you made my day."

I shrugged and continued. "We've talked a lot about the advances in computer chips, but software development alone has made the thirty-two-byte PC—those are the more advanced models—ninety times more cost-effective than a mainframe."

"You really believe in rubbing it in, don't you?" Maury asked.

"You wanted good news. Sadly, though, the good news is very piecemeal. Much of what's going on with U.S. business and industry has a flavor-of-the-month quality to it. There are trendy concepts floating around. Management will try 'empowerment,' or add 'teamwork,' then 're-engineering,' and don't forget the ever popular 'total quality management' approach. They are all great concepts, but the way most arc implemented breeds a 'This too shall pass' mentality, rather than lasting, fundamental change.

"It helps to explain why the Fortune 500 has given way to the Foundering 500."

Children and Rules

"Last time we considered the question of when it's time to fold 'em. Now let's focus on something that isn't quite as drastic: calling a spade a spade. By that I mean recognizing that there is a problem. An honest appraisal, no matter how painful, is the very first step toward a solution.

"The profit-and-loss sheet at Glenn's company may look pretty good—and it does—so how can he identify the symptoms of a disease that may be in its early stages?"

Lonnie raised his hand. "I think what you said about price competition in my case is right on. If Glenn is competing primarily on the basis of price, he's got trouble."

"Absolutely. Price competition is a fool's game. The death of a thousand cuts. And one good way to tell that you're playing a fool's game is when your competition starts to proliferate. When they see a weak spot develop, they'll pile on. Anytime a new competitor appears, start worrying. The vultures are circling.

"This is why price competition is deadly: To hold your prices steady or to reduce them to expand market share, as we saw with Lonnie, you start eliminating resources that are vital to an organization's long-term survival.

"And it works the same way with individuals. To compete with Third World workers on that basis, Americans will have to drastically cut their life-style. What happens then?"

"It's happening now—never mind *then*," Doug replied.

"Doug's right," I said. "Here's the scenario. My income declines, or doesn't keep pace. I get a second job and work twice as hard. However, the stress is killing me. The quality of my work slips; my health and morale deteriorate; my employers, alarmed by declining productivity, order layoffs and demand more output and contract concessions; I can't send my children to college; my old car breaks down constantly; I'm late for work twice a week; the government's tax revenues fall off, and it starts printing money to pay the bills, which fuels inflation. And on, and on, and on.

"You can see where this is heading—straight downhill. All because I'm competing on the basis of price with a Mexican, a Taiwanese, or an Indonesian worker. Wait until the Chinese get their act together! We can't win."

"What are the options, Dan?" Samantha asked, obviously sobered by the grim prospect that I had just laid out.

"Redefine competition. Have you ever watched children playing? If the game isn't turning out the right way, they'll just change the rules. Adults have to learn from kids. If the game hinges on price, we've got to change the rules and redefine the basis of competition.

"There are many different strategies or combinations of strategies to use as your primary competitive weapon. The key is to redefine the competitive strategy, but do it—most importantly—in the eyes of the customer. Here are some alternatives to price."

Alternative Types of Competition

1) Price-based ✓
2) Reputation-based
3) Image-based
4) Time-based
5) Innovation-based
6) Values-based
7) Service-based

"We have already talked about price-based competition," I pointed out. "The U.S. airline industry always comes to mind as a great example of the lose-lose nature of this strategy. Almost the entire industry has been at the edge of bankruptcy since it began to focus on a price-based competitive strategy (otherwise known as price wars).

"Reputation-based and image-based strategies are also old standbys. IBM is an example. It effectively combined an image-based strategy with a reputation-based strategy as the central focus throughout the highly competitive 1980s. Notice where I put 'service-based' on the flip chart. I placed it at the bottom because it provides the foundation for the rest. Without it, the others are greatly weakened. That was where IBM went wrong. IBM's past success was indeed its worst enemy. The corporation was too big to notice that many of its competitors began to redefine what service meant in the eyes of the customer. Its reputation was weakened and the company's image fell in the early 1990s. Belatedly, its new service-based competitive strategy provides a strong foundation to support the company's effort to enhance its image and reputation.

"Lonnie is going to be using innovation-based competition—I hope all of you are. The new technological tools give us a decisive competitive advantage over price—if—if we know enough about the tools to use them to change the rules of the game.

"This is the weapon the Japanese have used with such devastating effect. They *had* to innovate because they lost World War II. Their business and economic infrastructure was destroyed. One of our country's biggest problems is that we never blew ourselves up. Our factories and infrastructure were intact. As good as . . . I can't say new. It was as good as old by the time the 1970s rolled around. We didn't feel the pressure to innovate because everything worked, even if it was inefficient."

"*If it isn't broken why fix it,*" Maury said, rolling his eyes.

"Let's just say innovation was not a top priority. Now what happens when your competition is playing the innovation game, and you are playing the price game?" I asked. "You're doubly vulnerable. You've denied yourself the advantages of innovation, and you've fallen into a trap. Thirty-five years ago, Japanese labor costs were lower than ours. Innovation leveraged that advantage and forced the United States to choose between cutting quality or cutting costs—which inevitably impacts on quality anyway—just to stay in the game.

"Well, we didn't stay in the game very long. If you're still wondering why the United States is having trouble competing in the 1990s, take a look at this next chart. The game was all but over in many respects back in 1987."

Born in the USA

U.S. PRODUCERS DOMESTIC MARKET SHARE (%)

	1970	1975	1980	1987
Color TV	90	80	60	10
Videotape recorders	10	10	1	1
Phonographs	90	80	60	10
Machine tools	99	97	79	35
Audiotape recorders	40	10	10	0

"The irony is that we are masters of creating new tools of technology. We dream up this stuff, invent it from scratch. The Japanese knew they couldn't compete with our research labs, so they didn't. They focused instead on innovation-based competition by creatively applying our technology to create new products and make a killing. There are dozens of stories about inventors who begged U.S. manufacturers to try their stuff but got nowhere until they took the ideas to Japan."

"That's one reason we should have export curbs on critical technology," Glenn said.

"Wrong! We're talking about intellectual property—ideas," I said. "You cannot successfully erect barriers to keep ideas inside the national boundaries of any country. When you try, it stifles innovation. The tendency then is to try to lock up the precious new technology to keep it out of circulation.

"The U.S. music industry, basking in the profits of music CD sales, was appalled when, in 1991, Sony introduced a CD-quality digital audiotape recorder (DAT), instead of holding back on the technology to give it a chance to milk its cash cow for several more years. The very thought of rendering your own cash cow obsolete after only a few years makes the skin crawl.

"But that mind-set was the one that led to inaction throughout the 1950s, '60s, '70s, and '80s. The assumption was that innovation could wait, could be put on hold, while the capital investments made decades earlier were being recaptured several times over."

"Is there any law against getting a return on your investment?" Maury asked gruffly.

"No, but letting your cash cow die of old age in the meantime probably constitutes cruelty to animals.

"Let's go back to the second type of competition—time. I've shown you examples of how Du Pont, Alcoa, and Corning have been leveraging time—and profits—with technology. When you examine the figures, the advantages of speed are obvious. I like a comment made by John Young, the CEO of Hewlett-Packard: 'Doing it fast forces you to do it right the first time.'

"What Mr. Young is suggesting is that if you don't set aside time for fixing mistakes caused by the process, more productive effort will be put into successfully integrating the design and manufacturing systems, thereby reducing costly mistakes. The proof is in the profits. Consider this: High-tech products that come to market six months late but on budget will earn 33 percent less profit over five years. Coming out on time but 50 percent over budget cuts profits only 4 percent.*

"People will actually buy time. In several former Iron Curtain countries, used cars now sell for much more than new cars. Why? New cars take months, and in some cases years, to get after the order. Used cars can be delivered immediately, and that's worth twice as much.

"Time-based competition actually provides you with a new product to sell. Maury thinks he is in the mainframe business, but if he developed a reputation for getting orders out ahead of schedule, for providing maintenance ahead of schedule, and for launching new models ahead of schedule, time would become his company's major product. And that's a product that will synergize with any other product."

"That kind of reputation has got to give a company tremendous flexibility," Lonnie observed. "It could basically move into any area and be regarded as a major force in the market even if it's relatively inexperienced with a particular product line."

"Now you've explained why the arrival of the Japanese makes everyone shake in their boots," Brenda said. "If Toyota announced tomorrow that it was planning to enter the hospital business in the United States, nobody would laugh and say 'You're a carmaker, what do you know about hospitals?'"

A Sign of Decline

"Dan, aren't you identifying a symptom of a troubled organization when you talk about speed or lack thereof?" Glenn asked. "It seems to me that if you could track a company's speed in delivering its product, and if there is significant deceleration, there's a major problem setting in."

Doug picked up on Glenn's point. "Among other things, that deceleration is confirming that the company is not using time-based competition and is probably bogging down with a cost-cutting strategy."

"You're both seeing the benefits of time-based competition," I said. "It goes right back to the rule I gave you several nights ago: *Learn to fail fast.* I'd rather not spend $50 million and five years only to discover that I was digging my grave. Much better to drop a few million over six months, recognize the mistake, and start over.

*An economic model prepared by the consulting firm of McKinsey & Co.

"One problem, though, with time-based competition," I added, "is that the concept has been so popularized that most of your competitors already know all about the benefits. A time-based competitive strategy, therefore, isn't unique and doesn't provide the edge it once did. It's important to keep the basics of each rule evolving by creatively applying new technology. The rule, *Find out what your competition is doing, and do something else,* really comes into play."

Great Expectations

"Dan, you said that the seventh type of competition, service-based, is the foundation to build the others on. Why is the service economy being called the wave of the future?" Samantha asked. "Isn't service-based competition flying in the face of the need for new technology?"

"What I'm referring to when I use the term service-based is not fast-food chains, although McDonald's now employs more workers than GM and, therefore, the service economy is a force to be reckoned with. I'm talking about providing a 'product' that exceeds the customer's expectations."

"Exceeds expectations . . . ?" Lonnie asked. "What's the point of that? Customers aren't going to buy things they don't want."

"Want to bet? Here again, the Japanese have used service-based competition against us. At one time American car buyers expected small cars to be chintzy, uncomfortable, unreliable, and underpowered. The Japanese auto companies deliberately set out to exceed those expectations. Starting in the mid-1970s, the consumer who bought a Toyota or Datsun because it had the lowest price tag got what he or she wanted—cheap transportation. But in the bargain, consumers also got what they would have wanted, if only they had known they could get it."

I paused. "Sounds familiar, doesn't it?"

"It's one of the rule cards," Samantha said.

I nodded. "Consumers got what they paid for along with several extras—comfort, reliability, and power. The Japanese exceeded the customer's expectations. And what else?"

"They ate Detroit's lunch," Lonnie replied.

"That's for sure, but I was looking for another rule. Can anybody supply it?" Silence. "Okay, I've used this phrase before but haven't dealt it out as a rule card. Let's get it on the table."

I continued, "Quite a few years ago, the chairman of Sony wanted to introduce the Walkman to the American market. The company did a U.S. survey and found out that hardly anybody was interested in a little portable cassette player with headphones. Most executives would have said 'Well, that's that . . . forget the Walkman.' But Sony went ahead anyway and changed the way people thought about a little portable cassette player with headphones. It became a personal stereo system. It became an urban oasis of pleasure and solitude. It became an indispensable time-saver. Not a little portable cassette player with headphones.

"Recently, Marriott changed the way I think about morning coffee. I normally call room service the night before and order coffee for 6:00 A.M. the next morning. If the coffee arrived five to ten minutes late, I considered room service to be satisfactory. In the last ten Hilton hotels I have stayed in, that was the time range. However, in the last ten Marriott hotels I have stayed in, the coffee arrived exactly when I asked, to the minute. I was not only satisfied, I was impressed. I have changed the way I think. Now I expect that kind of service from all hotels. By using time-based competition, some new tools of technology to coordinate, track, and schedule the orders, and then tying it to customer service, Marriott changed the way I think and made the competition look bad.

"How do I know if I'm exceeding the customer's expectations? In Hilton's case, they may not even know that Marriott changed the way the customer thinks. Remember, you need feedback before it's too late.

"A good friend of mine, Michael LeBoeuf, wrote an excellent book a few years ago titled *The Greatest Management Principle in the World*. The essence of what he was telling the reader was . . . " I went over to the flip chart and wrote:

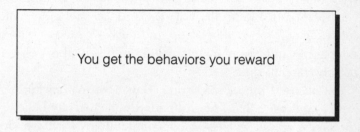

You get the behaviors you reward

"It's important to reward feedback. Develop a system of rewards for customers who go out of their way to help you get better. This is another example of opportunity management. Crisis managers wait for complaints or sales that fall below expectations, loss of market share, and internal talk of mature markets, down cycles, and industry slumps."

Doing Good, Doing Well

I moved back to the flip chart. "Personally, my favorite type of competition is values-based competition. It's what makes me stop at the neighborhood store where the owner 'forgets' to charge the poor family from down the street for a few items each week. It's what makes me go to see every Walt Disney movie that comes to town. It's what prompts a growing percentage of people to buy gasoline from ARCO—remember those whales several years ago that ARCO helped free from the Alaskan ice? Many people liked that.

"Values-based competition uses values as the main strategic weapon—loyalty, honesty, openness, understanding, generosity. Values-based competition involves doing what your intuition tells you is the *right* thing to do, rather than the bottom-line thing to do. It's hard to measure returns from values-based competitive practices; therefore, you don't try. The focus is to make a positive difference above making a profit. This tends to be a highly profitable competitive strategy."

I turned to Maury. "Don't snicker. This form of competition used to be as American as apple pie. Some of our greatest institutions were founded on it. Once upon a time, many of our buying decisions were based on a very personal reading of the vendor's honesty, openness, loyalty, and generosity. It's why people get so outraged when a big company says there isn't a problem with their product when all of their customers know that there is a problem. Or when a local large manufacturer cuts blue-collar salaries and, at the same time, raises executives' pay, or closes up shop and moves overseas without helping the people and community transition into something else. It's not a neighborly thing to do."

"You accused me of snickering, Dan. I wasn't snickering," Maury insisted. I could see that I had offended him. "I really believe in what you're saying about values-based competition. We have donated a large number of PCs to inner-city school systems, even while our sales were falling, to the point that I've had insurrections on my board of directors. What goes around, comes around. If you're unfair and disloyal to the workers and the community in which you operate, that's the treatment you'll get in return."

There was only one thing to do—apologize. "Sorry, Maury, I guess I'm too used to the Attila the Crumb school of management. Maury is right on. These values really do pay off. Example: Mazda's plant at Flat Rock, Michigan, near Detroit, employs 3,100 and the UAW represents the workers. Unlike other plants, it uses only two worker classifications: production and maintenance. This allows Mazda flexibility in moving people around to various jobs, and there's an efficiency premium that's not available to other U.S. automakers. In return, the workers get security. Mazda pledges 80 percent of regular pay during layoffs.

"Would Mazda have won that concession without being generous and concerned about its workers? Maybe, maybe not. But without give and take, when it's all take and no give, I can guarantee that the organization will suffer competitively: poor morale, high absenteeism, low productivity, deteriorating quality.

"Another example: Southwest Airlines has a no-layoff policy and a profit-sharing plan. Employees own 13 percent of the airline. The policy encourages workers to be committed to the firm's long-term future."

"Dan, if Maury is using values-based competition, why is his company having trouble?" Glenn asked.

"For one thing, price-based competition still takes precedence—" I started to say in reply, but Maury cut me off.

"I can answer that. I've got a mature cash cow, I'm not seeing the new big picture based on the new technology, I haven't embraced change—shall I continue?"

"No, I follow you," Glenn said. "It's not enough to just change the basis of your competition."

"Exactly, Glenn," I said. "There are seven things to look for when calling a spade a spade."

Are You Seeing . . . ?

- Mature markets
- Mature products
- Rising costs
- Growing number of competitors
- More customer complaints
- Increasing price-based competition
- Orders falling short of expectations as the new norm

Avoiding the Big Squeeze

"This talk about competition has made me think that it could be my hospital's biggest stumbling block," Brenda said. "I don't know *who* we are competing against, and therefore I don't know *how* we're competing."

"That's another good way to determine whether there is a problem—another way to call a spade a spade," I said. "There's an old poker maxim that really applies in this case. When you sit down to play for the first time with strangers, look around the table for the sucker. If you don't see a sucker, get up and leave because you're the sucker.

"Look around for the competition, find out who they are and what they are doing. If you draw a blank—hello sucker!"

"I must be one then because I can't come up with any competitors," Brenda said, shaking her head. "Sure, there are other hospitals in the area, but we're not really competing with them the way Maury competes with IBM. There's more than enough business to go around."

"Why are hospitals closing then?" Doug asked.

"They're being squeezed out by higher costs, increasing numbers of uninsured patients causing most emergency rooms—which have high operating costs—to lose money. And when that's not a factor, out in rural areas, for instance, dwindling population is the kiss of death," Brenda replied.

"But the hospitals that survive are the ones that have been effective competitors in that they can absorb the effects of the higher costs and the shrinking patient base," I said. "They are competing successfully because they are not losing as fast as the others."

"I understand that. Still, our survival as an institution does not depend on what the Sisters of Mercy Hospital across town does or doesn't do. If Glenn doesn't sell enough Maxwell House, he's out of business. If a patient goes to the Sisters, we're not likely to fold."

"I don't handle Maxwell House," Glenn reminded Brenda.

"Anyway, you get what I mean," she said.

"What would happen if the patient didn't go to Sisters of Mercy? What if all their patients came to you?" I asked.

"We'd be swamped. It would be a nightmare."

Samantha chuckled. "That's like me saying 'No thanks, I don't want all those policyholders from Northwestern Mutual. I'd be swamped.'" The chuckle gave way to a hearty laugh. "Please don't make me write all those policies!"

Brenda reached over and gave Samantha a friendly shove. "Okay, okay," she said.

"Your hospital and the Sisters of Mercy Hospital are both caught in the same rising tide of health care costs," I said. "By as much as 15 to 20 percent a year. Your survival depends on generating enough revenue to pay those bills, meet your payroll and earn a profit. Every patient that crosses the threshold of Sisters of Mercy Hospital puts you in danger."

"Unless the patient is indigent," Brenda said. "Then it's costing them money."

"Values-based competition. If the health care industry thinks it has political and public relations problems now, just wait until all private hospitals shut their emergency rooms and bar poor people," I said.

"Good point. I guess we are in competition but don't like to admit it."

"Acknowledging that competition exists doesn't mean that you have to conduct yourself like a predator. In fact, the new rules go a long way toward making predators extinct. Health care, like any other industry, should play this rule card."

I pointed out, "One of the factors driving health care costs higher and higher is a variation on the old keeping-up-with-the-Joneses obsession. When one hospital gets a new piece of the latest technology, every other medical center in town wants one. Much of this expensive, high-end technology could be shared."

Brenda nodded and reached for the pack of new tool cards. She shuffled through it but without finding what she was looking for. "I was going to draw the fiber-optics card," she said.

"Fiber optics is one of the twenty core technologies," I said. "What did you have in mind?"

"Fiber optics would be the perfect way to link all the hospitals in the city. Not only could we share systems, but we could transfer information back and forth and centralize many of our functions like administration, insurance claims, and the storage of medical records. It would be great!"

"The place to start with fiber optics is right in your own hospital. Link every nurse's station, every desk, every patient's room with fiber optics," I suggested. "That hospital of yours is choking on information, and fiber optics can carry 100 times more data than conventional methods. *Network with all* of the various specialty areas, from pharmaceutical and admitting, to the operating room and the patient's room, to X-ray."

"My goal is to become the chief information officer of the hospital, which, until we started these games, I thought was about as realistic as aspiring to be captain of the Titanic. But some of these tools make it very doable."

"Some?" I asked.

"All," Brenda said, correcting herself. "Fiber optics is a core technology, but it can be used to link this card." She went back to the pack and immediately located what she was after.

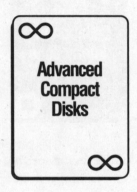

"I'm going to solve our biggest problem—information collection, storage, and retrieval. We've got many separate islands of information, whole archipelagos of information that don't have any contact with the rest of the hospital. I'm going to tie them all together with a fiber-optic network and enter the Communication Age."

"What's that old saying," Lonnie asked, "be careful what you wish for because you may get it? If fiber optics can carry 100 times more data, you're going to have 100 times more information roaring down on you like an avalanche."

"Combining on-line information with the transportability of CDs," Brenda said, "you could integrate the entire hospital staff, regardless of location. The day's lunch menu, prescriptions, medicare reimbursement schedules, doctors' telephone numbers, even X rays could be accessible to all.

"Each patient will get a bar code on his or her wrist bracelet. Zap, zap, and a full record will instantly appear on the screen. We could even give each patient a small, three-inch recordable CD that could hold all of their entire insurance and medical history, including compressed X-ray files."

"I guess you're going to play with this card then?" I asked her.

"Better believe it. When you first dealt out the cards I said 'That one's for me.' I'm going to go on the warpath against paper. When I'm through, we'll have the first paperless hospital. From prescription forms to luncheon menus, it will all be digital."

"Lots of luck," Maury said. "These computers and other high-tech things breed paper. There's more of it now than ever."

"You're right, Maury, but that's because we're still using methods and techniques from an age gone by," I said. "We've got one foot in the Communication Age and one foot in the Information Age. Here's an example of that kind of ridiculous straddle. The navy has smart bombs and dumb filing cabinets. A 9,600-ton navy Ticonderoga-class cruiser carries twenty-six tons of manuals on how to maintain and operate its complex weapons systems. The weight actually slows its movement through the water. All of it could go onto CD.

"The air force is just as bad. The B-1 bomber has one billion pages of technical documents. And it now takes the air force an average of twenty months to review and distribute the 2 million pages of maintenance updates that contractors send out each year on other weapons systems. Again, all of it could go digital."

"I still think it's ultimately impossible to go entirely paperless," Maury insisted. "Brenda said she would digitize everything including the lunch menu. On the first day the orderlies will print out copies, and every patient in the hospital will have a piece of paper on their bedside tables. It's human nature, and paper begets paper." Maury leaned back in his chair and folded his arms across his chest.

Brenda wasn't buying it, however. "Deal 'em, Dan," she ordered.

"I think the deck is stacked," Maury said.

"Maybe that should be a new rule card—always stack the deck," I said.

Brenda was getting up a head of steam. She was ready with an explanation of how she was going to use the card. "Forget those printed lunch menus; the orderly can show the menu to patients on his electronic notepad and let them check the items they want. It will be transmitted over the hospital's wireless radio system right to the kitchen with the name and room number. Wait a minute, I just got a better idea! Every patient gets an electronic notepad. If they want the day's menu, they can call it up on the screen, ask for an orderly to refill their water pitcher, or play a computer game to pass the time. They could even use it to write letters. Every hospital employee could also have an electronic notepad. We might even want to loan or rent them to hospital visitors and outpatients. No more visitors wandering around lost; they could call up a floor plan on the electronic notepad."

"One of the reasons that paper is still with us in a big way is that computer and digital imaging technology are just now reaching the price flash point, allowing a large number of us to make the digital commitment. Because of the graphic, user-friendly interface today's computers have, larger numbers of nontechnical people will discover that technology has given us the ability to handle digital information in a way that's more convenient than paper—and convenience is the key," I said. "Paper, for all of its nuisance potential when it piles up on the desk, is pretty handy. It's cheap; you can fold a sheet of it up and put it into your pocket. Information retrieval is a snap. A manual isn't a bad way to store information. Everything you need is between two covers. But twenty-six tons of manuals? That's not convenient, that's crazy.

"If I have an electronic notepad, the only reason I need paper is that there is no interface for the data that I'm dealing with. Why write it on paper when I can write to the electronic notepad with a pen-based input device? Or a voice-based input? If the data is already on paper, I'll enter it into my electronic notepad by scanning it with a hand-held scanner. The hospital will probably hire an outside company to scan the massive amounts of paper information we already have. From that point on, we will be digital to begin with.

"The electronic notepad is going to be more ubiquitous than the pager. Many people simply won't leave home without them." I gestured toward Brenda. "Eventually, the hospital won't need to supply visitors with electronic notepads; they'll bring their own and switch over to the hospital's wireless system within the building.

"The electronic notepad is so revolutionary because it gets the data out to where the work is being done. If you're on the go and standing while you work, an electronic notepad with pen input will be the way to go. When you are at your desk, a keyboard will interface with it easily for faster data entry.

"Good example: I was in a Lexus dealership a while ago and got a look at a beautiful high-tech service operation. There were several computer terminals linked by satellite to a national database and an advanced expert system. Only one problem—a big one: The computers were nowhere near the service bays. The mechanics had to stop work and go into this pristine environment to use the computers.

"You know what? They didn't bother. The dealer told me the mechanics disliked using the system. They preferred paper manuals. This new technology will change that. With electronic notepads and flat-panel displays, the computer will be as useful and familiar as the mechanic's wrench."

Frame Up

"I can literally play with every tool card in the deck," Brenda said. "And—"

"Stop right there," I ordered. "Repeat what you just said."

"I can play with any card in the deck."

"You're the first person who has said that, and you get the prize." I went to the hall closet and got her a three-by-four-foot picture frame on which I had mounted all the new tool and rule cards, faceup, like a huge poster. "Hang this in a very visible place for all to see," I suggested. "It's essential that we know about all the new cards and essential that we always consider playing with all of them as we face our problems and create new opportunities."

Brenda nodded and dealt herself a card from the deck on the table.

"We can lower inventory costs and increase our customer service by linking up to all the drug companies, the medical product suppliers, the laundry, the caterer—you name it," Brenda said and reached for another card.

"Can we use this one—or what? It's going to impact heavily on the art and science of diagnostics. There's going to be a significant speed and quality advantage. We should come up with our own expert system built around the hospital's particular talent pool and market it to other medical centers."

"Don't stop there," I said. "What about medicare?"

Brenda shook her head. "Medicare . . . I hate the word medicare. The paperwork is so brutal; we are not getting reimbursed for much of what we do. If anything puts us out of business, it's going to be that." She paused. "I don't see how an expert system is going to help there, though."

"How about an EDI linkup with the insurance companies, the government, and you. Let's put an expert system on-line that can guide you through the government red tape and is constantly being updated as regulations change," I suggested.

"Done. Give me another card."

Brenda continued, "Doctors are using cellular phones all the time to stay in contact. They even use them while they're making their rounds, and that's very expensive. The cost gets passed on to the patients in the form of higher hospital bills. PCNs will be much cheaper to use, and we can give them to nurses and the rest of the staff. No one will ever get out of range of voice communication."

"Patients too?" Lonnie asked.

"Patients too," Brenda replied as I dealt another card.

"Perfect for helping to train our medical students and educate our patients. Put a flat-panel in every patient's room for displaying digitized images like X rays and test results," Brenda said. "Next."

"Surgeons are going to love this!" Brenda held up the card. "They could simulate a difficult operation before they cut the patient; take a good long look at various organs; practice dealing with worst-case scenarios. And medical researchers will be able to use advanced simulations to conduct experiments that would take months and huge control groups."

Lonnie raised his hand. "Virtual reality. If I can tour the White Garden at Sissinghurst, Brenda's doctors will be able to stroll around the inside of my cranium."

"Don't laugh," I said. "The next time you get a headache it could be a visit from the faculty of the Harvard Medical School!" I turned back to Brenda. "You're doing very well. Let's see if I can trip up your momentum."

∞

Computer-Integrated Manufacturing

∞

"You did it, Dan," Brenda acknowledged.

"Computer-integrated manufacturing integrates every aspect of the manufacturing process from planning to production," I explained. "In some ways it is like EDI. All the bits and pieces are linked together. With CIM, if I shut off a machine for regular maintenance, the other equipment on the line is automatically adjusted accordingly. No more stagnant pools of information or half-finished products. The sales and marketing people would be advised, and the materials department would receive notification. Management will get an updated profit-and-loss projection for the quarter. Does this give you any ideas?"

Brenda nodded. "A form of CIM would work in our hospital. Once a diagnosis is made, the system would alert the appropriate departments. If it's a certain type of cancer, CIM would tell the pharmacy to prepare appropriate dosages for chemotherapy, radiology would be advised to schedule treatment time for the patient, family counseling would be notified to arrange a visit, and the kitchen would be told that a special diet is required for at least ten days."

"You're a star, Brenda," Samantha said.

"CIM is a card we are already using, and in some ways selling," Maury said. "Now some new applications have come to mind that didn't even occur to me until now."

"The rule is the one I introduced Tanya to last time: *Creatively apply technology*. Don't pigeonhole technology as just for hospitals or just for manufacturing plants."

"Where is Tanya, by the way?" Doug asked.

"She's at a Farm Bureau meeting. She called me to say that she was going to make a presentation on biotechnology."

"Good for her!" Maury said. "But that's something right there that proves you're wrong to say we should play with all the cards. How am I going to use this biotechnology stuff?"

"Biochips," I said.

"Bio-whats?" Maury asked.

"Biological chips. Research is under way on using biological molecules to build microelectronic components. Maury, you could be manufacturing a computer someday that is the size of one bacterium."

"I used to say 'impossible' to statements like that. From now on I will say 'improbable,'" Maury replied.

"I like that, Maury," I responded. "Improbable not only gives you an out if time proves you to be wrong, it also leaves you a little more open-minded."

Brenda spoke up. "Hit me with the rDNA card, please."

"Coming right up," I said.

"We all agree that medical cost containment is one of the country's biggest problems. Recombinant DNA, and genetic engineering in general, has great potential in the area of preventative medicine. If we can head off chronic illnesses before they even start by identifying genetic propensities and take early action in the form of life-style changes, the savings will be enormous." Brenda pointed to the deck. "There's another related card that I should have too, Dan."

I knew at once what she wanted.

"When Dan first mentioned it," Brenda explained, "I went back to the hospital and asked about antisense. This technology can block, neutralize, or switch off genes that express themselves in the form of life-threatening diseases. How much do you think we'll save if antisense can shut off genes involved in cancer?"

"You do that and there will be doctors looking for work," Doug said.

"Doug's point is an interesting one," I said. "We assume that tomorrow is going to be just like today. We have cancer as a major killer today, therefore we'll have cancer as a major killer tomorrow. If I'm involved in the health care industry, what happens if antisense comes along and shuts off genes vulnerable to cancer without the need for chemotherapy, radiation, and months of costly—and profitable—therapy?"

"Kiss your cash cow *adios*," Maury said, sounding more like Lonnie than a seven-figure CEO.

The Kindest Cuts

"Brenda, you've shown that you can use all the cards, so I'm not going to lay out the entire deck," I said. Brenda looked disappointed. "Okay, three more."

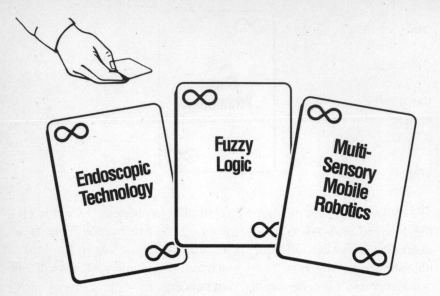

"Go for it!" Samantha said.

"I thought for sure you'd give me desktop videoconferencing," Brenda said. "That's going to be a real help to doctors and patients." She stopped and furrowed her brow. "But robots aren't bad either. Plenty of dirty work in a hospital for a robot to do. Any repetitive job like floor cleaning, laundry, food service, and rubbish removal could all be handled by robots. Patients could be monitored by a stationary robot in each room. If his or her temperature rose or breathing patterns changed, for instance, the robot's sensors would detect the fluctuation and summon the nurse."

"I don't think I would want this big robot standing in the corner looking at me in my hospital bed," Samantha said.

"Keep in mind that a robot does not have to look like a robot; they're not confined to a limited number of sizes or shapes. Since the robot that Brenda just described doesn't have to move, a stationary computer with sensors connected to the patient and a communication channel to the nurses' station is all that would be needed. Remember my simple definition: Robots are basically computers that can move." Samantha looked relieved.

"Can I play?" Lonnie asked.

"Sure."

"The gurney that transports patients to surgery could be robotized," he said, "and then you could use fuzzy logic."

"I'm glad you took that one, Lonnie; I can't figure out a use for it," Brenda said.

"If a robot stops and starts with a jerking motion, it would dump the patient off the gurney the first time it had to stop. Fuzzy logic lets it slow down or speed up so gradually that the patient wouldn't notice it." Lonnie beamed. "No nasty surprises."

"I'm going to have fuzzy logic put on our elevators," Brenda vowed.

"Newer buildings are already starting to be set up with fuzzy logic elevators," Maury added. "They are so smooth that the only way you can tell that they have moved is when the door opens and you're on another floor."

"Last card," I said, holding it up.

"Easy. The endoscope is going to allow for what we call minimally invasive treatment. Why get cut wide open when a small incision will do the job? Joints, the abdomen, the brain can all be operated on with the endoscope, although it's called by different names depending on where it is used and according to the cutting tools that are involved."

"Can anybody else think of an application for the endoscope?"

Glenn gave it a shot. "It has the ability to transmit video pictures down the tube, right?"

"That's right."

"Might make a good security surveillance system in my warehouse. We had thieves get in there a few months ago, and they simply used a ladder to get up beside the camera and point it in another direction. The endoscope is so small they wouldn't know it was there."

"Spies would go for the endoscope," Lonnie suggested. "Sneak up and slip it under the door to see what's going on. Great for Peeping Toms, too."

"Only you would think of that, Lonnie," Samantha said.

It was time to wrap up the session. "You should all be pleased. You're really starting to change the way you think," I said. "Brenda, after running through all those cards, have you given any more thought to your long-term goals?"

"Maybe chief information officer is a short-term goal. I think I'd like the hospital to set up a medical network that will tie all the nation's hospitals together for education, diagnostics, and wellness programs. You said to network with all. I wouldn't mind running that. I could even open a business that specializes in helping medical centers go digital. It's a huge market. Another thing I could do is work with the government on cost-containment measures and work out the medicare payment nightmare. And then . . ."

9

Capitalizing on Creativity

Using the Trump Card

"I need your advice," I said to the group as we moved toward the dining room to begin another game. I paused to turn on the overhead light and let the others get seated. Most of them had staked out favorite places by then. Lonnie tended to sit on my right, Tanya on the left. And as usual, Maury Andrews took the power position directly opposite me at the other end of the table. "I've a friend named Paul who is in danger of losing his job. Let me tell you about him.

"Paul is an associate in a leading architectural firm. He's been there for about five years and has done very well. I first got to know him when he came to me for information about how new technology would affect his profession. We spent several hours talking about architecture and Paul provided me with a very telling insight into the rapid pace of change that was—and still is—taking place.

"He said, 'Dan, until fairly recently, maybe ten or fifteen years ago, the ghost of Leonardo da Vinci could have strolled into any architect's office in the world, sat down at the drawing board, picked up the tools of the trade, and gone right to work. He wouldn't have missed a beat.'

"Think of it—Leonardo was born in 1452 and died in 1519," I told the group.

"Paul went on to add that now, given the last ten or fifteen years of rapid change, da Vinci's ghost would be at a total loss. He'd look for his old tools on the drawing board and find a computer terminal and screen. Talk about an unhappy, irascible ghost!

"Almost 500 years of business as usual blown away in little more than a decade."

"And I thought the food distribution business was changing quickly," Glenn said.

"It is, and as quickly as architecture," I said. "You just don't have such a dramatic benchmark as Leonardo da Vinci."

I went back to Paul's story. "My friend's problem is this: He recognizes that rapid, fundamental change is taking place, but realizes that his fellow architects are still thinking like a bunch of 500-year-old men and women. I won't say that they're thinking like Leonardo, because he was probably the first twenty-first-century thinker way back in the fifteenth century.

"Basically, architects are trying to operate the new tools with the old mind-set, attitudes, and assumptions.

"Paul is in trouble because he is questioning everything—*everything*. The partners are going crazy. They don't mind buying new computers and plotters and programs, but they draw the line at using fifty elementary school kids as a design committee for the new school that they have a contract on.

"Paul refused to attend a meeting with the school board—the client—and spent the week in a kindergarten room crawling around on the floor playing with blocks and crayons.

"'Will you please meet with the client,' one of the partners begged him. 'Why? What do they know about schools? The last time they spent any real "fanny time" in a classroom was thirty years ago,' Paul argued.

"Meanwhile, the kids were cranking out drawings of the floor plan, using CAD software on Paul's notebook computer, choosing the locations of the bathrooms and the gym and the library, which, by the way, is three stories high and shaped like Mother Hubbard's shoe; Paul plans to build it out of a new plastic material that can simulate other surfaces, including granite, and it's stronger than steel and can be molded into any shape.

"The senior partner is aghast. His architectural signature is marble and granite. Plastic instead of granite?

"Paul compounded the offense by asking the kids to solve a problem the firm has with two other schools it built—using concrete blocks with granite and marble accents, of course. The decorative marble now is decorated with scratches, and the granite has been sprayed with graffiti. One nine-year-old came up with the graffiti solution while she was helping her mother wash dishes. She told Paul the next morning while they were kicking a soccer ball around the playground, 'You know that wall at the Beethoven school with the paint on it?'

"'Sure. It's got a huge skull and cross bones—Jolly Ludwig.'

"'Spray the wall with Teflon,' she said.

"Ten days later, Paul had Teflon on that wall, and the senior partner was furious."

"Did it work?" Maury asked.

"The coating covered up the original graffiti, and whenever anybody splashes new artwork on the wall, the school janitor washes it off with a hose," I said.

"I wonder if they could use a diamond thin-film coating to protect the marble from scratches?" Tanya said.

"Hey! That's my card, you can't play that!" Lonnie protested.

"Watch me. There are dozens of implements around a farm that could be coated with diamond thin-film to preserve a cutting edge or protect them from the weather."

I welcomed Tanya back—she had missed the last game—and asked how the Farm Bureau meeting had gone. She apologized for playing hooky. She said the meeting had been called on short notice and went on to give us a brief summary of what had happened. "They're a little leery of biotechnology. But I persuaded them to let me look into setting up an E-mail system. That wasn't quite as threatening. One of my first electronic letters is going to be about a new study that shows a higher-than-average cancer rate among farmers probably the result of their exposure to pesticides over a long period of time. I'll tell them about Monsanto's work on plants that don't need pesticides, and who knows, maybe the company will help me set up the E-mail system. I'll never find out unless I try."

"Sounds unethical to me," Maury said with more than a hint of sarcasm.

"Sounds like *networking with all* to me," Brenda added.

Nuisance Value

I resumed my story. "The idea of using a protective Teflon coating was such a good idea that the senior partner eventually convinced himself that he was the one who thought of it. By the way, using a diamond thin-film coating as a solution to the scratched marble problem was also a creative idea, Tanya. Current diamond thin-film application techniques don't lend themselves to this application, but there is always the future. You may want to follow up on that one.

"Despite his success with the Teflon, Paul still is on the boss's blacklist. Why? Robots. The kids love robots. One boy told Paul that it was silly to spend so much money building a school when robots would do it for free. And another idea was born!

"There is an assumption that most of us share—if we're adults—which is that only humans can build buildings. Paul contacted a company in Japan that has developed a robotized construction technique where the building basically builds itself. Starting after the foundation is laid, a module known as a superconstruction floor is put into place, consisting of vats of concrete, steel girders, prefabricated floors and walls, and various robotic cranes, welders, and concrete sprayers. Once the ground floor is complete, the superconstruction floor moves upward, climbing its own steel frame as it is put into place piece-by-piece.

"Paul had committed sacrilege! The architectural firm's principal construction company went ballistic about it and tried to have him fired. Paul, to get even, briefed the owners of a new construction company that is trying to compete in the city, and now they are looking into obtaining a robot crew. They've promised to use Paul as their architect.

"Paul, as far as the partners are concerned, has turned himself into a major nuisance. He disrupted a meeting with a prospective client just the other week by suggesting that the businessman, a former politician who has done very well as a real estate developer, consider the possibility of building his new mansion totally out of glass.

"At first, everyone thought Paul was being mean to the guy by bringing up the old line—people who live in glass houses shouldn't throw stones. He was a former politician, after all, and pretty good when it came to throwing stones.

"Actually, I think Paul got the idea precisely for that reason, but he wasn't making an editorial comment. The glass-house maxim gave him the inspiration: Why use wood or brick—or marble and granite? There's new glass being developed that is lighter and stronger than steel. And the funny thing is that the client loved the idea. It appealed to him as a wry joke, and a glass house was something no one else had. The senior partners have been trying to dissuade him ever since, and they may lose the job—all because of Paul!

"How can we help Paul?" I asked.

"He needs to rediscover the virtues of being a team player," Maury said.

"Who wants to play on a 500-year-old team that's on a losing streak," Doug responded.

"But he's hitting his head against a brick wall," Brenda said. "If he's not careful, he'll get fired."

"Maybe they'd be doing him a favor," Lonnie said.

"Unemployment is not a favor," Glenn said. "It sounds like the trouble I've had with my father accepting the new tools and rules. Dan suggested that I had to do a better job of selling the benefits of change. And he was right. Dad is starting to feel that he is gaining control of the business rather than losing it. Paul should probably try to persuade the senior partner that there are tangible benefits to be gained."

"As someone with no small ego," Maury said, "I can detect that the senior partner has . . . let's say . . . a high opinion of himself. Play to that. Give him a vision of his firm building twenty-first-century schools all across America. What Andrew Carnegie was to the municipal libraries in 1900, he'll be to the schools of the future."

"Good psychology, Maury," I said. "I liked the V-word: vision. Creativity usually starts with a mental image that soon becomes a vision. Paul has to get that guy's creative juices flowing. His—Paul's—are flowing just fine. But as a result he seems out of step with the rest of the firm. He's perceived as a pest.

"Paul realized that if we can rebuild our downtown shopping areas to attract people, we can rebuild our schools to attract students. When he started going to kindergarten class and getting down on the floor with the kids to play, he was reawakening his vision of what it'd be like to be a child. Adults make all kinds of decisions about kids, but they do it from the perspective of adults. Try looking at the problem of building a new school from the perspective of the customer—the child.

"How else are you going to solve the customer's future problems today? From an adult perspective, there's no problem here. What difference do bars on inner-city school windows make to kids?

"A big difference! So vision is very important. We've got to get a picture of the situation. That's why I've insisted that all of you come up with relatively detailed personal goals. In Paul's case, he could appeal to the senior partner's ego, as Maury suggested, this way: 'Sir, picture yourself cutting the ribbon with the mayor and the governor as the new school opens, and the scene repeats itself again and again until you've presided over ceremonies in all fifty states. Picture yourself and the other partners testifying to a Senate committee about the schools of the twenty-first century. Picture yourself briefing the president on the subject.'"

"Picture yourself on the French Riviera enjoying the profits," Samantha suggested.

"Now, I see why you told us earlier that a company's long-range plans should be reducible to pictures," Glenn said.

"Not reducible—expandable. Vision drives success because it drives creativity. One of the reasons that we are so poor at setting goals for ourselves—poor at allowing a vision to expand our horizons and capabilities—is that we are afraid of creativity. Scared stiff.

"Maury, play along with me on this. Here's a sentence: 'Marjorie is creative, I'm going to reward her.' Cross out the words 'creative' and 'reward'; replace them with two others that immediately pop into your head."

Maury nodded. "Marjorie is trouble, I'm going to can her."

That got some applause.

"Exactly. We know what happens to creative people," I said. "They get kissed off, and it starts happening early in life." I turned to Doug. "What happens to creative students?"

"We control them," Doug said with a straight face. "If that doesn't work, there's always '. . . 'Mrs. Mathews, your son is being disruptive. Perhaps you should speak to him.'"

"Which translates to 'Mrs. Mathews, you had better make Johnny's life miserable at home because he's making my life miserable at school,'" I suggested.

"Close," Doug said. " 'Mrs. Mathews, either you squash him or I'll do it for you.' "

"And that's why we're not going to play with the cards tonight," I announced. There was a murmur of disapproval. "We're going to work on some creativity techniques. My aim is to tap the spirit of Leonardo da Vinci that is within all of us."

"So that we'll end up like little Johnny Mathews or your architect friend," Glenn said. "Thanks."

I shook my head. "Vision. I want everyone here to envision the Mona Lisa."

"Smile," Lonnie said.

The Rewards of Silence

"I want you to think of the last time you spent a half hour just being creative," I requested.

After a minute or two, Samantha raised her hand. "When I picked out my Christmas cards."

"When I cooked a birthday dinner for my wife," Maury said, revealing one of his dark secrets: He loved to cook. There was some friendly ribbing, since the hobby didn't seem to fit his image.

"Every time I go to work," Lonnie said, "I'm an artist after all."

"Give us a break," Brenda said.

"When we were children we spent a good portion of every day just being creative," I said after it became obvious that the group had run out of examples of recent creative moments. "When we were kids, we made up stories, games, imaginary friends. We ran and jumped and rolled around for no other reason than it seemed like fun at the time. Some of our happiest memories of childhood grow directly out of that unstructured creative play. Psychologists call it 'free-play.' The child just makes it up as he or she goes along. Free-play is absolutely essential to that little person's intellectual and emotional development.

"All too soon, though, children start 'growing out' of their creativity. They really don't grow out of it at all. They are forced out of it by adults who make demands like 'Grow up!' and 'Act your age!' Tests show that a child's creativity plummets by 90 percent between ages five and seven. A significant amount of research indicates that by the age of forty most adults are about 2 percent as creative as they were at five.

"Remember what we talked briefly about the last time we were together?" I moved over to the flip chart and wrote:

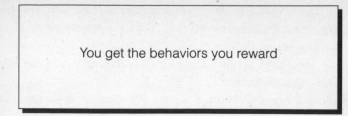

You get the behaviors you reward

"It's human nature," I said. "In fact, Pavlov and his dog demonstrated that all living creatures respond to rewards and punishment. Early in life we started getting rewards for being uncreative.

"Here's a line from the ancient Roman poet Horace."

"There is also a sure reward for faithful silence."

I continued, "He wrote that in 40 B.C., give or take a few years. In the interim centuries, there have been many rewards, and much adult silence. We've been bribed out of our natural and spontaneous creativity. Why?"

"Because Marjorie is trouble . . . ," Maury replied.

I nodded. "Luckily, Marjorie didn't grow out of her creativity; she will pay for that 'mistake' for the rest of her life. Give me little five-year-old Marge, five-year-old Ben, Suzy, Larry, and Linda—and I'll assemble a critical mass of creativity. If I'm clumsy, I get a chain reaction with the force of an atom bomb. Properly directed, there will be enormous creative energy.

"That's what my friend Paul was tapping into for the school building project. He assembled a pint-sized think tank to solve his problems. As a former teacher, I can assure you that kids can—and do—come up with creative solutions to dilemmas that totally baffle adults.

"I've taken community problems, nursing home problems, hospital problems, business problems, and carefully explained them to upper elementary school kids. Every time I do it and hear their answers, I have to say, 'Why didn't I think of that?!'

"There's a bonus, too. It's one that Doug will appreciate. The kids are far less disruptive in class because they have been given an outlet for their creativity. They are proud of themselves and gain self-confidence. They are not *silenced*.

"I have even seen fifth-graders come up with million-dollar patentable ideas, and they're actually getting them formally patented. That's right. They can do it. The kids are a tremendously underutilized resource. I would like to see a computer network started where we get our nation's kids on-line with each other—locally and nationally—and turn over some of our national problems and let them use their creativity. Maybe they could come up with some solutions; adults aren't doing too well with drug abuse, homelessness, and the deficit. I believe our nation's children are one of our biggest unused national and natural resources."

Tomorrow's Customer

Maury was taking notes furiously. He looked up and said, "I'm going to set up a kids board of directors." He chuckled. "Talk about an insurrection on the board; they'll try to lynch me." He tore a page off his yellow legal pad, folded it, and put it into his shirt pocket. "I've paid millions to think tanks for research. Some of the ideas were good, some of them were useless. We could underwrite think tanks in our elementary schools. . . ." He trailed off, intrigued by the various ramifications of the idea.

"A good example of values-based competition would be to donate or lend the school your next-generation mainframe to act as the center for a national student electronic problem-solving exchange network," I said. "There may even be an additional future benefit. Children have a way of growing up, and when they go out to buy a computer of their own, they might remember what you did for them."

Doug was equally excited. I could see it in his eyes. "Dan, business is totally missing the boat. If corporate and entrepreneurial America would make a major commitment to our schools, it would be a win-win situation for us all. The effect it would have on our nation and its future work force could be very positive." He shook his head. "Some companies are generous, but it astounds me that there are so few of them. You mentioned values-based competition and ARCO. That company quietly supports many public schools in the Los Angeles area, and there isn't much publicity about it. They're not alone; many companies are quietly involved in their communities, but unfortunately most businesses find it easier to fix the blame than the problem. Far too many mouthy advocates of free enterprise and foes of big government just stand back and expect government to take care of it."

"But do you want . . . Xerox or some other corporate giant educating your kids?" Lonnie asked.

"Why the hell not?" Doug and Maury answered in unison.

"Somebody's got to do it," Doug added.

The Music Within

"My original question to you was, when was the last time you spent a half hour being creative," I reminded them. "And I got some pretty lackluster answers. A half hour of creativity a week isn't much to ask. It could be creative play, structured or unstructured.

"My friend Paul gets his creative juices flowing by playing the drums. The thing is—he doesn't know how to play the drums. He just sits down and bangs away for a half hour. When he gets up from the drum set, he is raring to go. And, by the way, he's become a pretty good drummer but won't admit it.

"Making music is a great source of creative inspiration. I do it myself. Over the years, I've tried a variety of different instruments. Some I can handle, some I can't. Long ago, I realized that I'll never know unless I try. It's all part of life's learning experience. Pick up the violin and fool around and say, 'Well, now I know that I'm not cut out for the violin . . . I'll try something else.'

"It would be a shame if I were another Paganini but never knew I had the talent. I forget who said it—perhaps Albert Einstein—and I'm paraphrasing: What a tragedy to die with your music still trapped within you.

"The same thing goes for using your creativity."

Orchestration

"So much has been written about being an effective manager," I said. "The one bad thing about the word 'management' is that we tend to manage the things we don't like. That's why I cringe whenever I hear or read about 'managing change.' We see change as a problem that needs to be managed. What's being said is, 'We'd rather not change, therefore we'll *manage* change.'

"There's a similar problem when it comes to the term 'managing creativity.' It almost seems like an oxymoron. Back to old Marjorie: Marjorie is trouble, we'll manage her nagging, boat-rocking creativity by getting rid of her. Manage creativity by stamping it out wherever it rears its ugly head!

"We'll have to redefine the term *managing creativity*. I'd be happy to substitute *orchestrating creativity*. And before these games are over, I'm going to come back to orchestration as a vital concept. We have to learn to be orchestrators of creativity, innovation, and success.

"Orchestrating creativity has nothing to do with people sitting around a table brainstorming. The atmosphere is too rigid. Isn't this true? We almost never do anything important in our offices—that's where we're constrained into rational, rigid thinking.

"How many times have you heard—or said yourself—'I can't get any work done in the office!' I know many executives who do their best thinking at home, or in the car commuting. How about good old Bruce who goes on vacation and immediately starts bombarding the office with phone calls and faxes filled with new ideas?

"When your consciousness is in a state of relaxation, your subconscious mind—the seat of creativity—is freed to act.

"To orchestrate creativity, build a work environment that encourages people to break the barriers of their own self-imposed limitations. Research shows that less than 1 percent of your brain cells and less than 5 percent of the total area of your brain are involved in conscious operations. Set out to plug into the unconscious mind.

"Apple Computers does it by giving almost every building in its headquarters complex a specific theme. Meetings and conference rooms aren't identified by cold, impersonal numbers: In the 'Land of Oz' building, the conference rooms are named 'Dorothy' and 'Toto.' Apple's Management Information Systems Group has meeting rooms called 'Greed,' 'Envy,' 'Sloth,' 'Lust,' and the rest of the deadly sins.

"If I were setting up a planning operation for you, Glenn, I'd have a conference room named 'Tomorrow.' Another called 'The Day after Tomorrow,' and a third, 'The Day after the Day after Tomorrow.'

"Silly? Yes, it is. Fun? Of course. Childish? Absolutely."

Take It Personally

"One thing we all agree about children is, they are emotional."

I turned back to Doug. "What do we say to high-strung, emotional little Jenny, Doug?"

"You've got to learn to control your emotions, Jenny," he said without a pause.

"The mythical, snake-haired Gorgon turned men to stone with one glance. That single phrase—control your emotions—has turned millions of men and women to stone," I said. "Orchestrate creativity by looking for ways to get emotional about the problem, the situation, and the goal. Then, we can find a way out of the rational traps we're in. A trap, or a prison cell, has four walls. And every problem has four sides to it. That's why the solutions are so elusive. We're adept at arguing all the sides of the question and getting nowhere.

"A friend who is a journalist told me that he was inspired by a comment James Reston, the former columnist for *The New York Times,* included in one of his memoirs. Reston wrote that he was asked once about his obligation to *The New York Times.* He replied that he had no obligation to the *Times;* his obligation was to the newspaper's readers. That's taking it personally. And that's how year after year, Reston could throw himself into his work with enthusiasm and joy.

"I need to warn you, though, about emotion. There's a dark side. Attila the Hun knows all about emotion. In this case, the emotion is *fear.* It's the one emotion that is the enemy of creativity. Now, it is arguable that fear of failure, or fear of humiliation and death, are great motivators. Samuel Johnson said that the prospect of hanging in the morning does wonders to concentrate a man's mind. But that's not being creative. It's being an escape artist, a Houdini.

"Spreading and exploiting fear is the last refuge of the incompetent. Ultimately it offers destruction, not sanctuary.

"Instead of relying on fear of failure, encourage fear of success. Past success, after all, is your worst enemy. I'm so fearful and suspicious of success that I keep creating new challenges for myself. I can never really succeed because I'm constantly redefining success as it relates to me. I've got another hurdle to get over, and another and another.

"Fear of failure creates a defensiveness that is the enemy of passionate, creative work. Thomas Edison failed 10,000 times as he struggled to invent the light bulb, but he considered it to be a useful exercise. He was learning all of the different ways it didn't work. Contrast that with—three strikes and you're out, Tom!

"Who is going to risk taking a chance, being different, thinking differently, rejecting favorite assumptions, when one is certain of being punished for being wrong?

"Apple Computers, again, has the right idea. Don't give out rewards for solving problems, give out awards for finding problems.

"Ask what's wrong, not what's right. How can you solve a problem if you don't know that something is wrong in the first place?

"Problem finders are the Saint Bernards of the corporate world. Big, lovable mutts with tremendous powers of observation. Only in most companies they don't get a bone for finding problems; they get the blame for stirring up trouble.

"We love to slay the messenger who brings the bad news. So who wants to be the messenger? Lonnie, how about you?"

"I don't do messages," he said. "I agree with Sam Goldwyn, the movie mogul: If you want to send a message, hire Western Union."

"That's one good reason for hiring consultants," Maury commented. "You can blame them for the bad news."

"Now, you've just written my job description, Maury," I said. "The bearer of bad tidings. Any number of times, I've gone into companies that have experienced a crisis and found people—it may be blue-collar production workers or white-collar engineers—who say in all honesty, 'I could have told them that would happen.' Why didn't they? Management wouldn't have listened, or they feared that a superior would write them off as a grumbler or 'a poor team player.'"

I walked back to the flip chart and tapped the page. "We get the behavior we reward.

"When I was a teenager my father got me a summer job at Allis-Chalmers, where he worked. Dad wanted me to get my hands dirty and experience the real world. The company had a policy of paying $300 for usable suggestions from the workers. I said 'Great. I've always been good at coming up with ideas.' But, on my first day on the job, my new fellow workers on the shop floor told me to forget it. 'Don't tell 'em anything,' I was advised. Why? 'Barney over there on the overhead crane gave them a terrific idea last year, but the boss said it wouldn't work. You know what? They went right ahead and used the idea without paying for it.'

"After a few experiences like that the suggestion box was always empty. Barney and his friends were using their creativity on the job, however. Emotion is the primary fuel for creativity. They were using their creativity to sabotage the company and get even. Barney's artful jockeying with the overhead crane could bring work to a standstill for hours on the shop floor.

"We get the behaviors we reward."

Think Small

"Creativity has nothing to do with IQ, or intelligence, let alone a college degree," I said. "Edison had three months of formal schooling.

"Here's a well-accepted list of personality attributes that are associated with creative individuals. Notice that few of them are commonly regarded as signs of great intellect or brainpower."

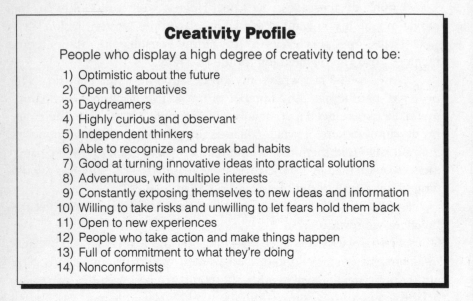

Creativity Profile

People who display a high degree of creativity tend to be:

1) Optimistic about the future
2) Open to alternatives
3) Daydreamers
4) Highly curious and observant
5) Independent thinkers
6) Able to recognize and break bad habits
7) Good at turning innovative ideas into practical solutions
8) Adventurous, with multiple interests
9) Constantly exposing themselves to new ideas and information
10) Willing to take risks and unwilling to let fears hold them back
11) Open to new experiences
12) People who take action and make things happen
13) Full of commitment to what they're doing
14) Nonconformists

I continued, "What I like most of all about the list is that I don't know anyone who looks at it and says, 'Not one of those characteristics fits me.' Creativity is not an 'all-or-nothing' thing. By deliberately collecting and cultivating a constellation of creative attitudes and habits, we can thaw out our frozen creative instincts.

"Run down the checklist to see how many you can claim as your own. If there's only one, consider it as a beachhead into the creative territory that you aim to capture. Work to increase the number. And when there's little or no progress, look around to see why. Are you discouraging your own creativity? Is your company, or coworkers and colleagues, rewarding you for lack of creativity?

"The National Science Foundation did an interesting study a few years ago. It wanted to determine the effect that company size has on innovation."

Company Size and Creativity

Firms with fewer than 5,000 employees:

1) Provided 4 times as many major innovations per R&D dollar as midsized firms.
2) Had 24 times as many innovations as firms with over 10,000 employees.

Conclusion = Larger firms use their size to resist innovation

---

I pointed to the final line on the chart. "That conclusion was not stated in the report, but it seems obvious to anyone who has worked in a large bureaucratic corporation. The larger the organization, the more it stifles creativity, primarily because of its almost inevitable vertical, hierarchical structure, which prevents the necessary orchestration—there's that word again—of creative talent. A symphony orchestra usually has a little over 100 musicians. More than that, and it becomes unwieldy. The orchestration literally suffers. Five thousand employees seems to be the break point for companies. I suspect it may be a lot less."

There was an audible sigh from Maury. "Just what I don't need—more downsizing!"

"Don't downsize—resize," I said. "There's a popular term making the rounds these days—right-size—and I like it because it implies flexibility. A good example of a large corporation that strives for the right size is 3M. It thinks big by acting small; creativity is encouraged. Workers are allowed to spend 15 percent of their time on whatever product-related projects interest them. One guy used his 15 percent to figure out a way to keep the bookmark in his church hymnal from falling out. He had an idea, and under 3M's rules convinced colleagues from other technical areas—marketing, sales, engineering, and so on—to join him in forming an action committee.

"It's a fascinating process, and it goes like this: The committee's job is to figure out how to design, produce, and market the idea. If necessary, the committee can apply to a 3M panel of scientists and engineers for a 'genius grant' to fund the work. As the project goes from hurdle to hurdle, the committee members are rewarded with promotions and raises. When sales climb to $5 million, to give you a for-instance, the product's creator becomes a product manager; at $20 million to $30 million, a department manager; and, at the $75 million range, he or she moves up to a division manager. Scientists who prefer to steer clear of management can join a separate track.

"How weird can you get, though? Some guy wants to keep his hymnal open to 'Rock of Ages' and 3M says 'Go for it.'" I turned to Maury. "If you told that to your board of directors, it would be golden parachute time."

"Tin-plated parachute time," Maury said.

I nodded. "And they wouldn't be raking in big bucks from a product called Post-it notepads. It's a $300 million business.

"Large returns within a context that's kept small and personal. Human touches, like division managers at 3M must know the first names of their employees," I said.

"Ouch!" Maury shook his head in amazement. "I don't even know the first names of all my division managers."

Lonnie couldn't resist. He wrote his name in large block letters on his notepad and held it up: "Hello, my name is Lonnie, but you can call me Mr. V-P."

"You asked for that one, Maury," I said.

"I guess."

"Back to the story of 3M," I said. "To ensure that this level of creativity is an ongoing affair, 3M requires its divisions to generate 25 percent of their revenue from products that have been introduced within the last five years. Instead of shunning the creative oddballs—getting rid of old Marjorie—3M executives go out of their way to find the most creative people, as though their success depended on it. And you know what? It does."

Do-It-Yourself

"'The fault, dear Brutus,' as Shakespeare wrote, 'is not in our stars'—make that corporations—'but in ourselves.' We play mental games and block our own creativity. In the 1980s, I was sharing with audiences a list of what I called creativity padlocks. One day, a friend gave me a great little book called *A Whack on the Side of the Head* by Roger von Oech. The book contained a list of blocks to creativity. Here's a combined list of generally accepted mental padlocks, things we say and do to lock away our own creativity."

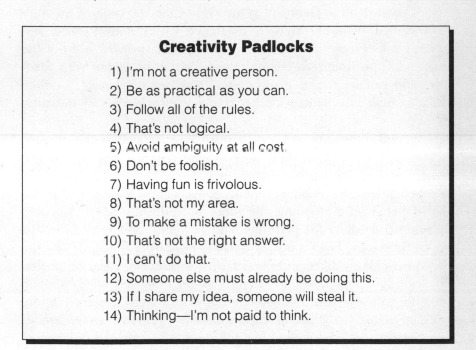

Creativity Padlocks

1) I'm not a creative person.
2) Be as practical as you can.
3) Follow all of the rules.
4) That's not logical.
5) Avoid ambiguity at all cost.
6) Don't be foolish.
7) Having fun is frivolous.
8) That's not my area.
9) To make a mistake is wrong.
10) That's not the right answer.
11) I can't do that.
12) Someone else must already be doing this.
13) If I share my idea, someone will steal it.
14) Thinking—I'm not paid to think.

I continued, "As soon as I saw these blocks to creativity as a metaphor for a list of locks, I started developing the keys. I've come up with five keys for opening the padlocks." I turned back to the flip chart.

Keys to Unlock Your Creativity

1) Observation
2) Incubation
3) Intuition
4) Emotion
5) Stimulation

"Consider them one at a time.

"*Observation:* In Germany at the end of the last century, an engineer named Wilhelm Maybach examined a perfume atomizer and noticed it had more than a nice smell—it had a process. He put the process of the atomizer—mixing liquid with air—together with gasoline, and came up with a carburetor. Maybach's associate, Gottlieb Daimler, built a totally new kind of transportation around it—the automobile.

"Spend some time by taking a fresh look at familiar objects. Ask, 'What haven't I noticed before about this thing?' Many of us look but we fail to see. The perfume atomizer is just a perfume atomizer. Try combining two seemingly different products or functions. I've found that the stranger the pairing, the more powerful the result. Maybach substituted gasoline for perfume. The new tool and rule cards that we've been using beg for this kind of *power-pairing*.

"*Incubation:* Immerse yourself in the problem, idea, or dream, consciously seeking a result. Then, back off from deliberate attention in order to allow your thoughts to mix with other elements that are present below the level of consciousness. We have an enormous capacity to store information, but getting at it on demand is a human frailty. Under hypnosis, a person could be brought into this dining room for just a few minutes and then later could provide dozens of details about the furnishings, what we were wearing, the pattern of scratches on the tabletop, Samantha's shade of nail polish; minutiae that the same person would forget once outside the hypnotized state. Yet, it is being absorbed unconsciously.

"Left to incubate, the vast reserves of information we have accumulated throughout our lives can surface, usually when we least expect it. When needed information does surface, stop what you're doing and write it down before it slips back into your subconscious. A Hungarian physicist who first conceived of the method for splitting the nucleus of an atom got the idea while he was crossing a London street. He was waiting for the traffic light to turn green, and when it flashed, the scientific principle that led to the atomic bomb was conceived.

"*Intuition:* People who use their intuition draw conclusions based on what they perceive as a pattern. The ability to recognize these patterns is directly related to the breadth and depth of their experience. Those who have the greatest exposure to a variety of experiences are better able to discern meaningful patterns.

"*Emotion:* The stronger the emotion, the more we can generate new creative solutions. Think of all the creativity that goes into an argument with a spouse or sibling. Many songwriters have written their best lyrics during a profoundly emotional time. Listening to music that causes an emotional response can be a good way to trigger creative problem solving.

"Finally, *stimulation:* The most gifted creators are constantly exchanging information and ideas with others. Studies have shown that scientists and engineers who talk most with their peers hold more patents, publish more papers, and produce more innovative work than their more aloof colleagues.

"Be a collector—a collector of ideas and information. For example, a corporate consultant and good friend of mine, Nido Qubein, said, 'Don't adopt what I say, adapt it.' I put this statement in my mental file with other related, powerful, and meaningful information. You could even put your collection on display for others to admire just like baseball cards or stamps. They'll offer criticism, and you'll gain a new perspective. Feedback and constructive criticism protect you from the creative person's biggest trap: falling in love with your own ideas. This often leads to an inflexible view of a concept that might need to be refined. Keep an open mind and withhold judgment until one morning when you step out of the shower and say—'Aha!!!'"

Metaphorically Speaking

"By comparing creative thinking to collecting baseball cards, I provided you with a metaphor. Cultivate your ability to make metaphors. When we say 'It's like . . .' we are building bridges, making connections that link seemingly unrelated ideas or situations.

"A metaphor is a visible crystallization of whatever it is that we are trying to come to terms with, which brings us straight back to the concept of visualization. I told you to visualize your plans for the future, to visualize your goals. Creative people visualize their ideas. Marconi saw information flying through the air and created radio. Einstein saw himself riding on a beam of light, and came up with $E = MC^2$. And it was Einstein who once said, 'The genius has mastered the art of mental imagery.'

"One way to get the mental imagery rolling is to ask yourself a series of 'what if' questions." I went to the flip chart and wrote:

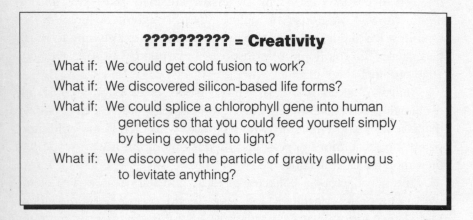

?????????? = Creativity

What if: We could get cold fusion to work?

What if: We discovered silicon-based life forms?

What if: We could splice a chlorophyll gene into human
           genetics so that you could feed yourself simply
           by being exposed to light?

What if: We discovered the particle of gravity allowing us
           to levitate anything?

"What if . . . I could figure out a way to save the hospital millions by heating the complex with thermal energy?" Brenda wondered aloud.

"What if I could figure out a way to instantly desalinate seawater right at the tap?" Lonnie asked.

"What if . . . " Maury paused as he peered out the front window. "What if I could figure out a way to design a tire that wouldn't go flat!" All heads swiveled in the direction of the CEO's limousine. Sure enough, the right front tire was riding on its rim.

There was general merriment around the table. Glenn volunteered to go to his car for a can of aerosol spray that would reinflate the tire and temporarily patch the hole. Doug suggested that Maury should just take the plates off and abandon the limo. "That's what happens to junkers down by the school," he said.

Maury shook his head and grinned. "You people are just full of great ideas."

"I believe a U.S. tire manufacturer has recently developed a tire that won't go flat, Maury, but it was a good 'what if' try," I said. "Let's go outside and see what we can do to be Good Samaritans, but first let me recommend that you all start compiling what-if lists. It's a good exercise that will keep stretching your imaginations. And I'll give you one to start the ball rolling." I turned the page of the flip chart.

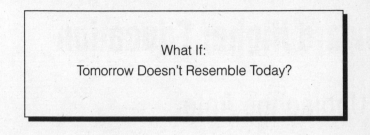

What If:

Tomorrow Doesn't Resemble Today?

10

Toward Higher Education

Upping the Ante

Doug Stedman phoned to tell me that he intended to drop out of the games because they seemed to pertain to businesspeople, not teachers. I was at the desk in my study when the call came in. I listened as Doug explained.

"I feel very frustrated," he said. "One minute I'm excited about the new tools and rules, and the next minute I'm down on them. Re-inventing the way we do business is one thing, I guess, but I just can't help but think that it's too late for me and too late for public education."

"I'd be inclined to agree with you," I said, "if the situation weren't so fluid and changeable. The new technology has broken all the icons, all the locks, all the guarantees. There's a level playing field out there. Today's losers—and as a nation we are losing—can be tomorrow's winners. Success today does not assure success tomorrow—as the United States can attest. We sat on our laurels for a good twenty years, and we're paying the price."

"By the time educators truly wake up, I'll be playing shuffleboard in a retirement village somewhere."

"Don't count on it, Doug," I said. "Retirement is about to be re-invented. If we do it right, it will become 're-engagement'; if we do it wrong, that golden-oldies village of yours is going to have all the charm and comfort of a *gulag*."

"I've only got thirteen more years."

"That's just about all the time Ted Turner needed to pull the rug out from under the three invincible money machines, the television networks. He started in 1980, and twelve years later the 1992 presidential elections were being described as the first postnetwork campaign. CBS and NBC correspondents, who were used to chartering Lear jets at the drop of a hat, wondered whether their accountants would allow them to pay for a seat on the White House press charter flights." I leaned back in my desk chair. "There are going to be a lot of changes in the next ten or fifteen years. In 1988, the Berlin Wall still stood, and the Soviet Union was a force to be reckoned with. And now?"

"Maybe we're headed in the same direction."

"You are pessimistic!" I said.

"It's been a bad day. I've normally got thirty-five kids in a class—which is too many—but I was running an average of around eight students today. It's like throwing a party and nobody shows. This absentee rate is driving me batty. These kids are never going to escape from the inner city if they don't start coming to school."

"Doug?"

"Yes."

"Do you have your pack of rule cards there with you?"

"I think so . . . here it is."

"Absenteeism is your biggest problem?"

"At the moment, yes. Tomorrow it'll be lack of books and equipment or something else," he said.

"Spread the rule cards out and pick one that will cover absenteeism—your number one problem. I'll pick one on this end." Doug agreed.

"Okay, which one did you draw?" I asked. He told me, and I put his card on the surface of my desk.

I also laid out the card I had selected.

"You've got an average of eight students who show up for class," I said. "That's a start."

"But don't advise me to skip the other twenty-seven—that will really put me in a bad mood."

"No, skip trying to figure out a way to solve the absentee problem. Those eight kids are the solution. If you creatively apply technology, they will get so excited about school that word will spread and nobody will want to miss out on the fun and excitement."

"Dan, the kids who come to school are called nerds by the ones who stay away, who call themselves dudes."

"What do the dudes do all day?" I asked.

"Hang out, shoot hoops, go to the video arcade, watch television . . . things like that."

I looked through the pack of tool cards and selected one.

"How about introducing molecular physics to your nerds by using your classroom computers?"

"We only have a few computers in our school, Dan," Doug responded. "Technology is not distributed evenly from school to school or from classroom to classroom."

"Do you think you could use one of the computer science teacher's computers?" I asked. Doug indicated that he agreed. "The two of you, working with those kids and object-oriented programming, could come up with video computer games based on basketball and the principles of molecular physics. All the right ingredients are there: Your knowledge of physics, her skill with a computer, the kids' imaginations and love of basketball."

"What's that going to do about absenteeism? I've still got the eight students I started with."

"Open a computer video arcade in the school. Molecular physics on one machine, English composition on another—Grammarman! Math games, geography games, history. Make the dudes want to leave the street corners to have some fun and hang on to the money that they're blowing at the commercial arcade," I explained. "Fight back. The competition is beating you fair and square. The customers do not want what you're selling."

"A computer video arcade . . . " Doug trailed off into silence. "Dan," he finally said, "I just pulled out another rule card."

I glanced at my deck and played a hunch.

"Which one?" I asked, placing the card on the desk.

"Build a better . . . "

". . . path to the customer." I finished the sentence for him.

"How did you know?" Doug asked.

"I invented the game."

He laughed. "Anyway, there's an abandoned block of stores in the neighborhood near the school. We could put the arcade in there . . . take the product right to the customer instead of demanding that the customer come to us."

I was over a gigantic hurdle with Doug. It was the first time I had heard him referring to his students as "customers." Until that moment, the concept was too alien; it hadn't stuck even though Maury Andrews and I had used it during the first nights the games were held.

I asked Doug what had prompted him to choose the word.

"I think it was what you said in the last game about creativity and metaphors. Until then the whole idea of looking at education in a business context was repugnant to me. But as a metaphor, I can see where it might work. The more I thought about it, the more I realized that business is the central metaphor of American life and culture. If it's not business, it's sports. Both hinge on winning or losing and playing by a set of rules. By refusing to look at education in that way, I was refusing to play the game that nearly everyone else is playing. Maybe that's why public education is losing."

"I also said something about 'getting the behavior we reward,'" I reminded him. "The marketplace is reward-driven. We work long hours because we get a paycheck and, I hope, because it's fun. Consumers enjoy shopping and the products they purchase. It's money and time well spent. However, modern education has lost the capacity, in many cases, to provide sufficient and relevant rewards for the behavior it demands."

"I agree," Doug said. "We end up using coercion, the force of habit, the promise of delayed gratification—and is that a tough sell these days! Maybe the easiest thing to do is surrender and turn the asylum over to the inmates."

"So what's it going to be, Doug, thirteen more years of coercion, habit, delayed gratification in an instant-gratification world—or surrender?"

There was silence on the other end. "I'll see you tonight," he said.

The Blame Game

As it turned out, my reluctant poker-warrior was the first to arrive for that evening's game. I gave Doug a conspiratorial wink as I began a few minutes of introductory comments. "There's a variation of poker called 'take it or leave it.' Another name for it is 'shove 'em along,'" I said, putting my coffee cup down on the dining room table and picking up a deck of cards. "You play it by accepting or rejecting the card that has been dealt out. If you don't want it, the card gets shoved on to the player to your right. If he or she also rejects the card, it keeps moving to the next player and the next until—if no one accepts it—the card is discarded."

"Sounds like they should call it 'politicians' poker' or 'passing the buck,'" Lonnie suggested.

"I think you're onto something," I replied. "Doug is going to deal—and that's appropriate since the buck has been passed to teachers. He's going to be using a deck of cards that I've prepared specially for tonight's game. First, though, he is going to deal everyone a card, facedown, from the new tool deck and one from the new rule deck before switching to the other deck—a deck that approximates what he and his students face every day at school." I pushed the cards toward Doug. He shuffled and started to deal.

"These first two cards are your hole cards," I said. "They don't get shoved along, but I want to ask if anyone would discard if they could?"

Brenda peeked at her cards. "I probably would have chosen a couple of different ones," she said, "but I can live with these."

"The same for me," Tanya said. Everyone else agreed.

"Excellent. You're learning to play with all the cards," I said. "Okay, Doug, switch to your special deck and deal faceup."

The first card went to Maury.

He blushed slightly and pushed it toward Tanya. The card went to Lonnie, Glenn, Samantha, and Brenda. "Funny," I said, "nobody wants to play with the teenage pregnancy card." Doug dealt another card.

Maury grimaced and got rid of the card, as did Tanya and the others. "No takers," I said.

Doug kept the cards coming.

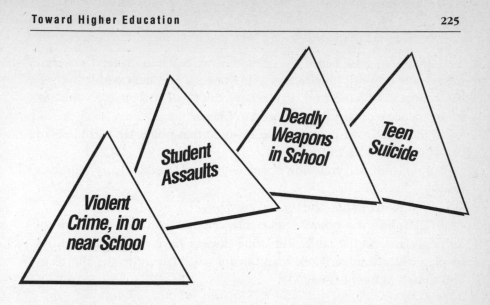

"You're probably wondering why these cards are triangular. Triangle-shaped cards are hard to shuffle and hard to deal, a fitting shape for cards with these face values." I gathered up the discards into a pile and evened off the edges. "What these cards have in common is that they were never dealt to any of you when you were in school. Although, Samantha is the youngest, and she probably experienced the leading edge of these problems."

"A couple of pregnancies and near misses," she said. "Not me—I might add—and beer parties."

"I wanted these cards to be out on the table because we need to appreciate what Doug and other teachers are facing every day," I said. "It is totally unprecedented in American history. Yet, teacher-bashing has become a national sport." I turned the page of the flip chart. "Take a look at these 1992 statistics."

Every School Day in 1992

- At least 100,000 students tote guns in school.
- 160,000 students skip classes because they fear physical harm.
- There are 40 students hurt or killed by firearms.
- 6,250 teachers are threatened with bodily injury.
- 260 teachers are physically assaulted.*

---

*National Education Association Statistics

"But, Dan, just based on performance, teachers haven't covered themselves in glory," Maury said. "If I have a salesman who can't sell the product, I blame him. If teachers can't sell their subject and its future value, who are we supposed to blame?"

I thought I was going to lose Doug at that point. He had the "I'm out of here" look in his eyes.

"I noticed that you didn't play with those cards, Maury," Brenda said.

"You didn't either," he replied curtly.

"That's just the point," I said, intervening. "You shoved the problem right around the table. But Doug doesn't have the luxury. He has to play with those cards each and every day." I went to the flip chart and turned to the next page.

Don't Fix the
Blame
Fix the Problem

"We're really good at fixing the blame. It probably goes back to our puritan overreliance on punishment rather than rewards as a motivator. But remember, I said that fear is the last refuge of the incompetent. Why punish the salesman for not making the sale when the problem is a product quality-control issue or a company that has developed a reputation for slow order fulfillment?" Maury crossed his legs and looked out the window. "Think of the valuable energy and resources that are diverted into hunting down scapegoats. We round up the usual suspects, and the usual problems continue to fester."

"I think what happens," Glenn suggested, "is that we get so wrought up about these problems like education that it short circuits the normal problem-solution dynamic. We alienate the ones closest to the problem and become stuck in the 'who got us into this mess' stage. Outrage becomes a substitute for action."

Lonnie raised his hand. "I also think it's the American way of dealing with failure. If there is a problem, some 'other' person caused it.

Once we accept the problems as *our problems,* we can begin to truly work together to solve them."

I held up two rule cards:

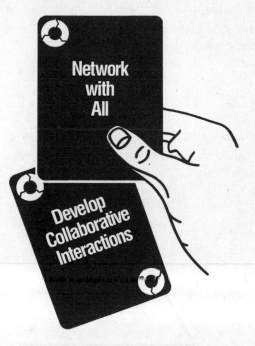

"Everyone thought these were pretty effective cards for business and hospitals and landscape design. But when it comes to education—forget it—the teachers are on their own. It's your fault, buddy!!! And by the way, after you finish teaching them math, science, English, and history, we want you to be a social worker, a counselor, a public health practitioner, a baby-sitter, a mentor, a surrogate parent, and a cop. But don't come to us bellyaching about lack of resources and support.

"We've got to learn to stop fixing the blame and to stand united in our determination to fix the problem. We need to network with all if we are going to fix a problem that involves us all.

"Why? Because we all share in the consequences of failure." I stopped and distributed Doug's discards to each of the players, tearing the drug and alcohol card in two. "I want everyone to have a piece of Doug's problems."

"I thought that if we shoved the cards along, they were discarded," Lonnie said.

"That's poker—this is life," I said. "No matter how many great new tools and rules you're playing with—high tech all the way—the *wild cards* in this special deck are still part of the game of education, and they are being played often and randomly. A winning hand is impossible for every player at the table if we don't eliminate these cards from the deck."

Bad Deal

I walked around the table and stood behind Doug's chair. "Doug has some other cards he wants to deal you, right Doug?" He just chuckled and dealt out a card.

Maury handed the card to Tanya. "Hold it right there," I said. "Tanya has had enough of obsolete equipment. Try playing with the cards that you were dealt, Maury."

"But I've got multimedia computers."

"If the status quo remains unchanged, our future work force—and that includes the men and women on your payroll—will know how to operate equipment that is obsolete by the time they get out of school, if you're lucky," I said. "And their reading skills will be marginal, at best, thanks to the large number of parents who never read a book, subliminally telling their children, if I don't value reading, why should you?"

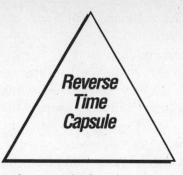

Maury looked warily at me before he passed the card to Tanya. "I thought I was lucky," she said. "When Doug dealt the tool cards, I drew the recombinant DNA card again. Now, I don't know."

"Walking into many of our schools today is like stepping into a time machine set to the past. The books are from the 1960s, '70s, and '80s, the filmstrips are from the '60s and '70s, the science equipment is from the early '60s, the building is from the '50s, '60s, or '70s, and, from the kids' perspective, the teachers are alive and well and living in the past. These are the kids who are our future, the workers and voters of the twenty-first century," I said. "After a steady diet of old filmstrips and textbooks, they're not going to know what to make of rDNA."

"Oh yes, they will," Lonnie said. He got up and went to the kitchen and returned with a dinner plate bearing a single vitamin pill. He put it in front of Glenn. "Would you prefer fresh ground pepper on your filet mignon, sir?"

"That's about it," I said. "Many people are so fearful of the new technology because of their lack of knowledge that the government will pander to them by making sure that the future is just like the present. And because many farmers like Tanya won't move from the business of farming to the business of biotechnology—they'll be moving toward the door marked Out of Business."

The card got as far as Glenn. "Which one of the rule cards do you have in the hole?" I asked him.

"Re-become an expert."

"Re-becoming an expert implies that there is a foundation level of expertise and experience to begin with—a starting point. Budget cuts are leaving our schools unable to even supply the basics." I asked him about his tool card.

"Advanced robots."

"Who's going to install, operate, and maintain them?"

"Another robot," Lonnie said—and nobody laughed.

Decade of Neglect

I moved over to the flip chart. "We're all sitting around scratching our heads in the 1990s wondering what happened. How did the most powerful nation on earth get in such a fix? You don't need to be a genius to answer the question. All you need is a memory, and not a very long one at that; just enough to remember the reports chronicling the 1980s. Here are some snapshots of the decade in the form of facts and statistics."

1983–1986

• On standardized tests, American students placed last in biology among 13 countries, including Hungary and Singapore, 11th in chemistry, and 9th in physics.

I turned the page.

1988

• Of the 3.8 million 18-year-old Americans, 700,000 left school without a diploma, and another 700,000 couldn't read their diplomas.

I turned the page.

- A National Science Foundation nationwide poll found that 21% of Americans think the sun revolves around the earth and 7% didn't know.
- Tests of 100,000 U.S. students in both public and private schools found that:

  Math: U.S. ranked 13 out of 15 internationally;

  Writing: 28% of high school students couldn't write a simple letter;

  Reading comprehension: 3rd-graders scored 38 out of a possible 100 and high school juniors scored 56;

  History: 54% of high school juniors passed; in literature 52% passed.

And to bring the 1980s to a close.

1989

- 77% of U.S. colleges give students a B.S. degree without their studying a foreign language; 45% without taking a course in English or American literature.

If you think the 1990s started out in better shape, look at this:

- The 1990–91 International Assessment of Educational Progress study found that U.S. 13-year-olds scored 13th out of 15 countries in science achievement but placed first in the amount of television watched each day.
- In 1992 the U.S. Department of Education released the following statistics:

  70% of U.S. high schools offer no earth or space science.

  30% of U.S. high schools offer no physics courses.

  17% of U.S. high schools offer no chemistry courses.

  7% of high school seniors are prepared for college-level science courses.

  The average high school dropout rate in the United States was 28.5%.

"So end the 1980s and early 1990s—and good riddance!" I announced. "You don't have to be a futurist to predict the future when there are statistics like that available. Seeing storm clouds on the horizon is easy when it's already raining and blowing up a gale. The strange thing, though, is that Americans are a hardheaded, practical people. I think that most outside observers of this country would have looked at the 1980s, at those same statistics, and said, 'Stand back— with a situation like this, Americans will do what they have always done: They'll figure out what's going on, get busy, and do something about it.'

"There was no need to stand back. The dropout rate continued to edge into the 30 percent range. Meanwhile, Japan managed to get 96 percent of its students through high school. Well, maybe a high school diploma doesn't count for much these days. But there is a frightening parallel statistic: With the growth rate of adult illiteracy—2.5 million new adult illiterates every year—it won't be long before the United States winds up having 30 percent of its adult population functionally illiterate. And you know what? We don't set very high standards for defining what it means to be literate. Anyone who has completed the equivalent of the fifth grade is considered literate in the United States."

"If I wanted to sound heartless, Dan, I'd remind you of what the Bible says: The poor are always with us," Glenn said.

"Thirty percent of us?" Lonnie asked. "There will be a lot more poor with that kind of increase in illiterate people each year."

"But there's the remaining 70 percent," Glenn argued, looking uncomfortable with his position. "That's a lot of hardworking, relatively well off people. And illiteracy doesn't necessarily equate with poverty."

Maury wasn't about to buy any of that. He rapped his knuckles on the tabletop. "If he can't read and can't write and can't count—and he's rich—I'd say his picture is probably hanging on the post office wall!"

"Adult illiteracy is a national tragedy," Doug said. "Nobody can claim that the illiterate are all criminals. But two out of three prisoners in our nation's jails are illiterate. You can't convince me that if our education system was functional, the only difference would be that we'd have jails full of prisoners with high school and college diplomas. I think there's a clear cause-and-effect relationship involved here."

"Another effect of illiteracy," I said, "is that it costs an estimated

$225 billion a year. Productivity suffers; we lose tax revenue, welfare and unemployment payments are a drain, and that's not counting the cost of prisons, police protection, and crime in general."

"If my customers could lick the shoplifting problem, their bottom lines would be a lot fatter," Glenn said.

"Again, as Doug pointed out, an illiterate person is not, *ipso facto,* a criminal," I said. "But what you're saying about the bottom line is worth a comment. In 1991, the American Association of Manufacturers did a survey of its members and found that 40 percent were having serious difficulty upgrading their technology. Why?" I asked, turning to the chart.

Partners in Decline

• 30% couldn't modernize because their employees were unable to learn new jobs.

• 25% couldn't improve product quality because of their workers' inability to understand new quality-control techniques.

"I call this chart 'Partners in Decline' because I think there is a direct correlation between our deteriorating educational system and the weakness of our economy. If 30 percent of our major manufacturing concerns are unable to modernize—and modernize they must— because workers are unable to learn new jobs, a huge portion of the economy is on the brink of disaster. We all know that these workers are *not* stupid and uneducable. They are coming to the job with an education that did not equip them for the 1990s. The future belongs to those who are capable of being retrained again and again. Think of it as periodically upgrading your human assets throughout your career. Let's face it, the corporate jewels are its information and its people, not its buildings and hardware. *Humans are infinitely upgradable,* but it does require an investment.

"James Rauch of the University of California at San Diego studied American cities with an eye to determining the impact that education has on the productivity of the work force. He found that for every year of additional schooling, productivity, on average, rose by nearly 3 percent.

"American business is engaged in a crash program to increase

productivity. That's a driving force behind organizational decentralization, as Maury Andrews will testify."

Maury nodded and said, "Chop, chop, chop. I've seen predictions
that downsizing will cost the economy 4 million new jobs by 1997."

"But what kind of Faustian bargain is that?" I asked. "It amounts to
high structural unemployment with catastrophic social consequences
in exchange for reduced overhead and minor productivity gains—usually without real growth.

"Here's my formula for increased productivity *with* growth and
without higher structural unemployment: new tools, new rules, and
new schools."

Maury pushed his chair back from the table. "You made a prediction earlier, Dan," he said. "Now, I will make one: If you bought junk
bonds or were into greenmail in the eighties, you're going to buy drastic downsizing in the nineties. We've decided that we're going to
destroy these companies to save them by hacking off huge chunks of
the work force and lines of business. And I'll admit it, I thought that
was the only way to go until I started looking at both of the new
decks. But you could downsize until you disappeared off the radar
screen, and without the new cards and workers who can play with
them and win, there's little hope. This economy is DOA if we don't do
something about the schools."

"DOA?" Tanya asked.

"Dead on arrival," the rest of us said in unison.

Teaching the Teacher

"Okay, Doug, where do we start treatment on the patient?" I asked.

"Quick results or slow?"

"Quick and slow," Glenn answered for me.

I told the group that I agreed with Glenn. "We should have *all*
been working on this problem—not just talking about it—in the 1980s.
The window of opportunity is beginning to close, and our international competition is going to slam it shut within the next five to eight
years. We will be too far behind to catch up."

"Five to eight years is real quick," Doug said. "First, teach the
teachers—all-out, full-bore retraining of our teachers. They have got to

re-become experts in education. And you know what? Most of them are going to enjoy the hell out of it."

"Simultaneously, we've got to start retraining the existing work force, too," Maury added. "It must be a two-track approach. Even if we waved a magic wand and fixed the schools tomorrow, I can't wait for the kindergarten class of 1993 to work its way through to college commencement. I've got a business to run right now."

I gave Maury the thumbs-up sign and walked to the flip chart. "Take a piece of paper, Maury, and write on it the percentage that your company spends on employee training," I suggested. He paused for a moment and then jotted a figure onto his notepad.

I turned and did the same on a blank page of the chart.

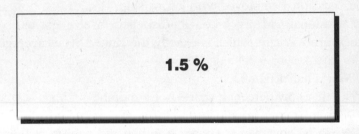

1.5 %

"Did I get it right, Maury?"

"You're high by a half percent, I'm ashamed to say."

I nodded. "According to the American Society for Training and Development (ASTD), U.S. corporations spend an estimated $30 billion on training annually. That's the good news. The bad news is that fewer than 10 percent of all corporations have any company training program. In a recent ASTD survey of 15,000 large- and medium-size employers, only 0.5 percent of the total account for 90 percent of the training dollars spent. We're going to have to start doing a lot better than that. One reason the figure is so low is that businesspeople don't want to spend money training the competition's workers."

"I'll give you a good example of that," Maury interjected. "Xerox has one of the best training programs around for its people. I go out of my way to raid Xerox for its best people whenever I can. I'm rarely disappointed with the quality."

"Nice guy," Lonnie commented.

"Tell me you wouldn't do the same thing," Maury replied.

"Does Xerox teach bonsai?" Lonnie asked.

"*Banzai*—yes," I said. "Xerox has actually won back market share from the Japanese after being hammered by them in the early 1980s. The emphasis on retraining and other reforms has paid off. Another success story: Motorola is pushing out 200 percent more pagers from a plant in Florida than it did in 1988 with only a 22 percent increase in the work force. How? Improved training."

"But what do you do about raiding?" Glenn asked. "I'm not Xerox or Motorola, I can't afford to be—in your words—'training the competition's workers.'"

"Training is an investment. If you treat it like one, and use accounting methods to track the return, you'd see some spectacular results—thirty to one in Motorola's case—that would more than pay for the occasional employee who jumps ship," I said.

"Occasional?" Maury asked in disbelief. "We've got a 4 percent monthly turnover rate, which is exactly the United States average."

"In Japan, it's 3.5 percent," I said.

"What's half a point?"

"No, that's 3.5 percent a year—not a month!"

Glenn took over from Maury: "Different culture."

"Yes, it is. However, there's a bonding mechanism at work that's universal. When an employer says to his workers 'I care enough about the company and about you to make an investment in your future success,' he is following a rule that we discussed the last time we met. Brenda, can you tell me what it is?"

"You get the behavior you reward."

"The employer who says 'Training is a bad investment because you'll be gone in a year or two' is encouraging here-today-gone-tomorrow habits," I said. "Maury said he was ashamed of that 1 percent training investment, and I'll bet we are the first people he's ever shared that figure with."

Maury nodded. "We don't hide it, but we don't advertise it either," he said.

"I believe that companies should be required by law to publish that figure," I said.

"It certainly would be a good quick way for job seekers to see where a prospective employer is coming from," Samantha said. "No offense, Maury, but I would hesitate to go to work for your company if I knew that's all you spent for training."

"I should have kept my mouth shut."

"Let's not pick on Maury," I said. "The point is that if the amount a company spent on training appeared in an annual report or a Securities and Exchange Commission filing, it would, at the very least, set some important standards. If we knew that Maury's competitors were outspending him in a big way on internal training, our judgment about his operation would certainly be altered."

"It would also make it easier for him to justify spending money for training just to match the competition," Tanya said.

"What, monkey see, monkey do in my business?" Maury replied with mock indignation. "Never!"

National Security

"Okay, Doug, we interrupted you with our discussion of retraining," I said. "Get us back on track."

"We need to re-invent teacher education before we can re-invent American education," he continued. "I'd like to see the federal government set up a national teachers' academy modeled after West Point and Annapolis. Congressional appointments, free tuition, rigorous standards, and when the students graduate they agree to serve a minimum of six years in a classroom or a school administrative position."

"Great, now you are moving toward redefining what constitutes national security," I said. "In the past, it meant well-trained military leaders and a heavy investment in military hardware. Now, and in the future, national security needs to include a well-trained and retrainable work force. It's worth an arsenal full of hydrogen bombs."

The idea of a national teachers' academy was warmly received by the group. Maury asked, "What would be the difference from the curriculum we teach teachers today?"

"In addition to teaching the new tools and rules, we need to cover subjects teachers are never exposed to because the schools of education did not consider business subjects relevant," I said.

"Give me examples, Dan," Maury asked.

I went to the flip chart and wrote:

New Basics for Re-becoming a Teacher
1) Advanced public speaking skills
2) Advanced listening skills
3) Nonverbal communication
4) Strategic planning
5) Managing creativity techniques
6) Personal counseling techniques
7) Personal development psychology
8) Knowledge of the new tools and rules
9) Time management
10) Total quality management
11) Customer service
12) Negotiations techniques
13) Sales techniques
14) Advertising techniques

"I referred to this addition to teacher education as a skeleton, and it really does work like one. Once these bones are in place, the musculature can be added," I said. "Whatever the teacher's specialty—math, science, history, English—these core skills are essential."

"What you've got there, Dan," Tanya said, "is a curriculum that you'd want any well-rounded person to be exposed to."

"Exactly. But sadly that's not what happens with our student-teachers. Does a teacher need to know how to sell?" I asked.

Doug answered, "Before I started to learn how to play this new game, I would have said no. As a teacher who now knows the new rules, my answer is yes. If I can't sell my students on that need to learn my subject, they probably won't."

"Most university administrations don't seem to understand that. The curriculum for their schools of education don't require them to take the subject of sales," I said. "Yet, I'll bet that terrific sales courses are being offered on the same campus. But there is very little cross communication between the various departments in a university. Intellectually and attitudinally, the school of education is far removed from the school of business."

"But some would-be teachers weren't cut out for sales. Wouldn't we be wasting a lot of teaching talent?" Glenn asked.

"To answer your question, Glenn, I'd like to ask Samantha about sales, since that's her area of expertise."

"Fire away," Samantha said.

"Can anyone become a great salesperson?" I asked.

"I don't think so," she replied. "There are certain personality characteristics that I think are needed for someone to be really great in sales, such as being outgoing and achievement oriented. Let's face it, in sales you can't allow yourself to get depressed over rejection; it's a daily occurrence. Not everyone could handle it."

"Now I've got a question for Doug," I said. "Do you think anyone could make a great teacher?"

Doug looked up at the ceiling as he thought for a minute. "No!" he said. "Some of my high school classes even send experienced substitute teachers home crying. It takes a special personality type to be a great teacher, although I suppose anyone can teach one-on-one under certain circumstances."

"When do you think college students who are majoring in education get their first experience in a classroom?" I asked.

Doug knew the answer and said, "Their senior year, when they student-teach."

"That's a nice time to find out that you hate kids or don't like discipline problems," Maury said. Doug nodded his head in agreement.

"You love being a teacher, don't you?" I asked.

Doug replied with a definite "Yes!"

"How would you like to teach English next year?" I asked.

Doug glanced at the floor as he said, "No thanks."

"As a teacher, do you know many colleagues who are teaching subjects they dislike?" I asked.

"Yes," he said. "Especially in elementary and junior high school. They are shuffled all around, since positions are filled based on seniority in the school system rather than on how much they love the subject they teach. If it's a matter of having a job or filing for unemployment, you take the position."

"Remember, we are not fixing the blame; instead, we are trying to identify the problems and find creative solutions," I said. "It seems to me that we have identified several solutions."

I went back to the flip chart and wrote:

Fix the Problem

1) Give undergraduate students enrolled to be teachers early experiences in the classroom.
2) Develop a personality profile of the best teachers.
3) Teachers need to identify their subject areas of passion and they should not be forced to teach out of those areas or face unemployment.

"Does a teacher need to know how to advertise?" I asked.

All of their heads started to nod yes.

I went on to say, "If MTV is an example of a teacher's competitor, it appears we'd better start learning how to build a demand for our product."

"The student is the teacher's customer along with the student's parents," Doug said. "A course in customer service seems like a winner."

I nodded. "The most effective customer service representatives make sure that the customer wins in some way. To accomplish that, he or she needs negotiating skills to ensure that the company's interests are in balance with the customer's desires.

"Does a teacher need to know how to negotiate?" I asked.

Again all of their heads nodded yes in unison, so I concluded by saying, "It appears to me that a teacher needs skills that are equivalent to those of any fast-track business executive. If my competition has the business skills of MTV, I'd better have the know-how that my competition brings to the job—otherwise he or she is going to keep on winning, and I'm going to keep on losing.

"So, Glenn, to answer your question, we're not squandering teaching talent by requiring business courses such as sales, we're enhancing it.

"Let's shift the focus from teacher training to students."

The New Basics

"Since I have kids of my own in school, the first thing that comes to my mind is the curriculum. Is it obsolete?" Lonnie asked.

"Our current general subjects are all valid. However, we need to define a set of objectives, some that are new and some that existed at one time but seem to have been overlooked or de-emphasized," I suggested.

The New Basics for Students

1) Ability to demonstrate adaptability in a rapidly changing environment
2) Ability to communicate orally
3) Ability to apply negotiating skills while demonstrating personal responsibility
4) Ability to work in collaboration with others (teamwork)
5) Ability to identify and apply the benefits of cultural diversity
6) Ability to identify and apply the benefits of being observant
7) Ability to identify and apply the benefits derived from service to others
8) Ability to focus and apply creativity in problem solving
9) Ability to demonstrate technological literacy in problem solving
10) Ability to apply computer technology to enhance task performance
11) Ability to find and communicate paper-based and digital information
12) Ability to apply memorization techniques
13) Ability to learn new skills and assimilate new ideas quickly
14) Ability to take initiative and be self-directed
15) Ability to apply abstract thinking techniques
16) Ability to identify problems and develop solutions

"If we keep these objectives in mind—and all of them are aimed at providing an education that is not frozen in time but upgradable in response to change—then there is great latitude when it comes to specific content. Any number of traditional courses fit into the framework. The crucial factor must always be relevance.

"Is learning to type a relevant skill, knowing that someday soon we will have voice input on our PCs and electronic notepads as the method of choice? No, not if it is only seen as a way to train someone to operate a current keyboard. Yet, if we look on it as a way to improve eye-hand coordination and as a high-speed data-entry technique that is faster and more efficient than handwriting, learning to type is relevant to a student's future."

"And the three Rs?" Lonnie asked. "Raging hormones, Riding around, and Resisting Mom and Dad."

"I'm a big fan of the three Rs—of the reading, 'riting, 'rithmetic variety," I said. "But for the twenty-first century let's add three Cs."

Old Three Rs	Plus	New Three Cs
Reading		Change
'riting		Creativity
'rithmetic		Communication

"Art, music, math, and science—all of the traditional subjects—will remain extremely important. Many subjects that teach interpersonal skills, such as sports and music, are subjects that are being cut back at the present time but shouldn't be. Foreign language skills can no longer be neglected. Two languages at minimum should be introduced when the children are in the lower elementary grades, when the ability to learn languages is at a peak, instead of waiting for junior or senior high when it has begun to atrophy. We must widen our children's horizons so they understand that the economy of the twenty-first century will be global in scope and they will be a part of it."

Superstar

"My turn to deal, Doug," I said. "But first I want to ask about your long-term goals."

"I don't want to leave the classroom. I guess that means I'm a little deficient in the vision department."

"Not really. It depends on what you mean by 'classroom.' Does it have to be *a* classroom with four walls and a ceiling?"

"No, but I'm not sure if I know what you're getting at," he said.

I dealt him his first card.

"Remember this one?" I asked.

"Sure do. The math teacher's best friend," Doug replied. "That's the one card I would have picked myself. In addition to doing the dirty work of teaching by rote memory the skills needed for math, it would be a wonderful way to introduce students to the ten greatest scientific breakthroughs in history. You mentioned Einstein wondering what it would be like to ride a beam of light. *You* could actually do that with multimedia computers."

"And you could ride that same beam of light," I said.

"How's that?"

"I'm thinking of this card."

"Compact disk technology is a key part of multimedia computers. And one of the core technologies is the laser. See where I'm headed?" I asked.

"Lasers and beams of light."

"Right. Ride it into classrooms all over the country. You're a tenth-grade science teacher, and at any given moment during the school day there are tens of thousands of tenth-grade science students hard at work. Until now one of the only ways to reach them was face-to-face, through a textbook, a filmstrip, or a video. The ten greatest scientific breakthroughs in history could be your first multimedia project."

"Me?"

"Why not? You're a good teacher," I said.

"I'm a good teacher, but that's got to be expensive."

Maury was the next to speak. "No, because my company is willing to pick up the tab for people with an action plan. Corporate America would like to help, but most of us don't know how. If you've got a good idea, we'll produce and market it."

There was backslapping and an exchange of high fives with Lonnie, probably a first for the CEO.

When the hubbub died down, I dealt another card.

"That's got to be another terrific teaching tool," Doug said. "And it seems to me that it would allow the school to save money. We could conduct some very elaborate experiments in class without breaking a single test tube or spilling any expensive chemicals."

"Also, the kids would love taking a walk through the human circulatory system with virtual reality," Lonnie suggested.

"Speaking of breaking test tubes, I had an idea for solving the school equipment-shortage problem by using optical storage, the enabling technology for advanced compact disks," Glenn said.

"I'll take it," Doug said, eagerly.

"Put bar codes on all that stuff—microscopes, beakers, books—and you will have a running inventory that will allow the science teacher at one school to swap the beakers she's not using for the microscopes she needs that are sitting in an empty classroom across town."

"Not bad," Doug said. "I always have plenty of what I don't need. With the bar-code reader and a computer, we could track our supplies like an industry using just-in-time inventory techniques."

"Just-in-time equipment for just-in-time learning," Tanya added.

"Link all the schools in the city with fiber optics—maybe the phone company would do it for free—and use it to exchange the bar-code data, files, and other teaching material," Glenn said. "I seem to remember reading about the phone company doing something like this in North Carolina already," he added.

"Heck, go for broke, the country is almost there anyway—link every school in the country with fiber optics and use digital imaging," Lonnie suggested. "Swap books, periodicals, course outlines, pictures, spread the wealth."

"And tap into the Library of Congress or the Smithsonian and use them as *national education utilities,* the way power companies plug into regional and national electrical grids," I said. The ideas were coming from every direction.

"Dan, I think Doug is another one who could use all the cards," Brenda said.

"I've got one that will stop him," Samantha predicted, reaching for the deck. She shuffled through it looking for the card and tossed it on the table.

"It stumped Brenda when you dealt it to her, Dan; let's see what Doug can do," Samantha said.

Doug furrowed his brow and then smiled. "Maybe I can't use it in my classroom, but I'm going to help the shop teacher set up CIM to give his students experience with totally integrating every step of the manufacturing process. The best students, who normally wouldn't be caught dead in a shop course, might discover that a computer is as important a tool as a drill press, or that there is more to it than just a glorified toy."

"Vocational education classes have always been offering applied science to students. Because it wasn't called applied science, the vocational education teacher and the science teacher and the guidance counselor somehow missed the connection. Most teachers would agree that all subjects relate to one another. Unfortunately, teachers have never forged a solid relationship between courses before, but now they should," I said. "Take all those isolated courses in math, science, health care, computer science, metalworking, English, small-business accounting, and apply our rule—*Network with all.*"

Doug pointed at me. "How about dealing me the interactive television card?"

"You've got it."

"I've been thinking about this one for a long time," Doug said. "Talk about networking with all, we can link up every school and classroom in the country. Our eleventh-grade algebra teacher handles the subject better than anyone I've ever known. With direct broadcast satellites or fiber optics, or even digital microwaves, interactive television capabilities could get to any school without having to wait a decade. She could be teaching algebra to every eleventh-grader from coast-to-coast."

"What I like about that idea," I said, "is that teachers who hate teaching algebra but do it because it's assigned can be freed up to pursue the subjects they really love. Technology can put passion back into the classroom, and your algebra teacher could be on the way to becoming a media superstar. Why do you have to be a rocker or a rap singer to be famous in this country?" I gestured toward Doug. "In that sense, technology will be liberating, but there is bad news."

"It figures."

"Teaching needs to become a full-time job before the end of the decade. Not twelve months in a classroom in front of the students, but full-time like any other professional who gets *paid* time to plan and learn. The teacher's job must include time for colleague-to-colleague meetings, research, curriculum planning, and retraining, with emphasis on retraining."

"Presumably that means the kids would go to school year-round?" Tanya inquired.

"Children need time out of the classroom but not all summer. Vacations should be spread out around the calendar. As it now stands, elementary school teachers spend September and October reteaching material that has been forgotten during June, July, and August. In effect we are losing over four months. What's worse, schools in the United States average about 180 days of instruction a year. Japan and some of our overseas rivals are pushing 240 days, and the school day lasts close to eight hours. By the time students reach the twelfth grade in the United States, their counterparts have gone an additional two years! Of course, our kids need the time off so they can harvest the crops. At least that's why it was set up that way in the first place. We're going to have to adjust our school year to the reality of our era rather than our grandparents' time."

"Are you kidding—240 days?" Lonnie asked. "My kids would go berserk."

"I think that's overdoing it," I said. "My preference would be 220 days, which would increase classroom time on the order of 10 percent. I'd also recommend a minimum of seven hours a day when school is in session."

"Doesn't a child's attention span start falling off after four or five hours, Dan?" Brenda asked.

"Watching their favorite television shows for four or five hours is not hard for kids to do, is it?" Brenda smiled, nodding her head in agreement. "It depends on how bored or engaged they are. Boredom has never been defined very well. Let me give it a try. There are three types of boredom. Either the subject is too difficult, or it is too easy, or you can't relate to it at all. Since it is hard to individualize instruction when the room has thirty kids in it, teachers have to aim for the middle. This causes boredom. With the new tool of multimedia computers, the students will be active participants in an individualized, entertaining, personalized way. This will free the teachers to spend more time with individuals who need human help, causing better relationships and less boredom.

"There is another factor we have to consider when addressing a longer school day, and that is the amount of comfort and mobility they have. If a child is immobilized at a desk for long periods of time, you're asking for trouble. Instead of cutting back on athletics and other beneficial activities, expand those programs to fill out the later hours of the day, when the students are starting to get restless. The New Basics for Students that I introduced you to earlier this evening would really come alive as part of physical education or sports programs. Another way to address this issue is to put air-conditioning and carpeting in schools. There are times when the heat and humidity are so high that no white-collar worker would put up with an environment like that. This investment would be well spent."

Doug shook his head. "I've already got enough to do teaching science, and the same goes for my colleagues."

"Of course you do," I said, "and so did I when I was in the classroom. More nonteaching support personnel need to be provided to free you up from all the supervisory chores that have been assigned to teachers over the years. A few good examples would be bus duty, hall duty, and the ever popular lunchroom duty. It would be a good thing to do with volunteer parents, or college freshmen and sophomores who are enrolled to be teachers, or students who are using government-sponsored loans to pay for their education, or retired professionals who are looking for something to do to keep them young."

Glenn cleared his throat. "And where's the money going to come from?" he asked.

"Everyone stand up," I requested. "Get out your purse or wallet and start digging. That's where the money is going to come from. Remember this rule card."

I continued, "People have already changed the way they think in the United States. And we'll all gladly dig deep and pay in the near future if we feel that our schools are taking good care of their customers. If the schools pull kids off the streets, away from the gangs and the guns and the drugs and the latchkey life, then society will be willing to pay for these services."

Doug stacked his cards in a neat pile. "Dan, if you had told me a couple of weeks ago that I'd have to give up my long summer vacations, I would have decided to bag the teaching profession. And that's a sad commentary when the best thing about your job is the vacation." He paused. "But not anymore. You're really talking about letting the teacher re-invent education from the inside out, and with public support, that's exciting. Having vacations like any other professional makes sense to me, especially if it would mean that I could make a salary comparable to other top full-time professionals. I wouldn't need the summer off—I wouldn't dream of taking the summer off and missing the fun—if I could participate in that kind of a revolution."

The group was still standing. I took a moment and looked into each face. "Normally, I don't like to be a prophet of doom, but here goes: By the end of the decade, education—for both children and adults—needs to be our biggest business. If it isn't, we will all be in big trouble."

11

Long-term Problems and Short-term Thinking

The Joker Is Wild

"Thanks to the process of elimination," I said, "it should be Samantha's turn to play." I glanced around the table and could see that she was definitely looking forward to the prospect.

"I was ready days ago," Samantha said.

"Good, but you're going to have to be patient just a little while longer."

"Patience is not one of my strongest attributes," she said, grimacing slightly.

"Wrong woman, Dan. Brenda's got patients," Lonnie said from the kitchen, where he was rummaging around in my refrigerator to find milk for his coffee. "Cancer patients, TB patients, AIDS patients . . . "

"Dan, would you mind seating Lonnie on top of the trapdoor tonight?" Glenn asked. "One more bad pun and away he goes."

Lonnie emerged from the kitchen, munching on a brownie that he happened to find stashed away behind the pickle and mayonnaise jars. "Have pity," he said, stuffing the brownie into his mouth and holding up his right hand with the fingernails toward us. Under the nails, there was a faint shadow of a day's hands-on work as a landscape designer. "Have pity for a man whose life is one of grime and punishment."

"Sit right here, Lonnie," I urged, pointing toward the head of the table. "The trapdoor is already open."

"And, for good measure, the chair is plugged into the nearest electrical outlet," Maury commented.

"Don't fry for me, Argentina," Lonnie said as he sat down.

Maury shook his head when the chair didn't start sizzling. "Darn . . . must have blown a fuse."

To get the group to refocus on the game, I did a showy, slow, hand-to-hand cascading shuffle with the cards. A dangerous stunt, since I end up spraying cards around the room every now and then. This time, though, I pulled it off. As I dealt out the cards, I suggested that everyone take a close look at the hands they were dealt.

"You've given us five cards, and one of them is the joker," Tanya observed.

"That's right. Anything else?"

Doug laughed. "Look whose face is on the card."

The card resembled one of those "Uncle Sam wants you" recruiting posters from World War I, but the face was Lonnie's.

"Aside from the remarkable resemblance to someone we know well, what can you tell me about the joker?" I asked.

"It's a wild card," Glenn said.

I nodded. "Under the rules of poker, the player who holds a joker can designate its value to round out a flush or a straight, or it can stand as an ace," I explained. "There is a joker in every deck, but I've given each of you a joker to play with. Seven different jokers around the table means there are seven different ways to louse up the game and to prevent a winning hand from emerging."

"I resent that," Lonnie said, looking at his likeness on the card.

"I couldn't resist having a friend of mine who is great at caricature put Lonnie's face on the cards; normally I use Uncle Sam or the Statue of Liberty. Why?" I asked. "Because the joker is government regulations and crisis management mentality.

"Like it or not, the government is slipping a wild card into each of our hands that determines who wins and who loses." I picked up another deck. "In fact, often it's an entire hand of wild cards, even an entire deck. Glenn, if you decided to set up a domestic satellite operation, here are the cards you'd be playing with."

The Licensing of a Typical Domestic Satellite Application

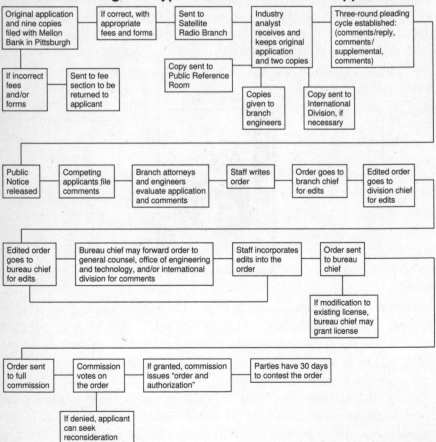

I continued, "Twenty-six wild cards. No wonder the domestic satellite industry has been painfully slow to get under way. What happens is that these administrative procedures are based on two outmoded assumptions: one, that a hierarchical organization is the only effective way to run a regulatory agency; and two, that paper—reams of paper—are necessary to execute decisions."

"I wouldn't even bother applying," Glenn said.

"That's a very understandable reaction," I said. "But that means you can't make use of direct broadcast satellite technology. The government is throwing similar wild cards in against many of the new tool cards, and this problem is going to get worse before there's any improvement. I hear a lot of talk these days about a business and government partnership. But, if Washington attempts to use procedures from the 1930s, '40s, and '50s to solve twenty-first-century problems, there's going to be a decline into social chaos."

"There's already chaos in the insurance business, Dan," Samantha said. "I call it government by land mine. Washington and the state governments are hemming us in with legislation and regulation. We're afraid to take a step for fear that what was a clear path last week will blow up in our faces today. It puts us in a totally defensive posture and deprives the public of improved products and services."

"Try filing for medicaid and medicare payments," Brenda said. "I think the regulations change from hour-to-hour."

Everyone had a horror story about the government. Glenn and Maury complained about the tax laws; Tanya was upset at farm subsidy payments that encourage overproduction; Lonnie was being driven crazy by the Superfund's rules for the disposal of toxic waste; and Doug was fed up with school aid cutbacks.

The gathering had turned into a general gripe session about the government. I let it go on for a while and then intervened. "What are we going to do about it?" The question stopped everybody cold.

Forging Links

Doug spoke up first. "As a teacher, I now realize that we're going to have to *build a better path to the customer* using the new tools and rules. The same goes for the government. Those wild cards you dealt out for a typical broadcast satellite application demonstrate how twisted and rocky the path has become."

"I'd start by improving the internal and external communication system," Maury said. "The government is going to have to enter the Communication Age in a big way."

"I agree, but let's begin at the beginning," I said. "Crisis management has become the government's modus operandi. Because of our four-year election cycle, long-term planning extends only as far as the next presidential race on the federal level. In addition, one- and two-year budget cycles interfere with continuity. As a result, every issue that's under consideration is white-hot, and getting a consensus is very time-consuming and difficult, if not impossible."

"But Dan, if we can't agree on how to put the fire out when the house is burning down around us, how in the world do we agree to long-term plans for the future?" Glenn asked.

"Ironically, once the planning process gets under way, it is easier to achieve consensus about the year 2020 than it is about next year. John Kennedy's vow to put a man on the moon was easy to swallow because it wasn't going to happen the next week or the next year. Take Maury's goal of entering the Communication Age. If that was laid out in *five- or ten-year increments,* the crisis managers who gridlock each other by battling over short-term solutions would say, 'Enter the Communication Age? Ten years? I'll buy that, but let's get back to more important things.' Planning and implementing the technological revolution could then unfold smoothly."

"Step number one," Tanya said, "is to start small with a voice-mail and an E-mail system like the one I'm setting up for the Farm Bureau. A government-wide, *intra-* and *interdepartmental* E-mail system would help eliminate stagnant pools of information by making sure that the right information got to the right people."

"Many government agencies have had technology such as voice-mail and E-mail for quite some time now," I said. "Access to technology isn't the main problem. Let's face it, all organizations, businesses, education, and government have access to basically the same technology for the most part. The question is, who is using it, and how is it being used? The answer is, few are really applying the technology effectively, and there is little or no integration and focus."

Tanya threw a card on the table.

Desktop
Video-
Conferencing

"I'll bet this could be used to heighten communications," Tanya said. "The ability to see each other might help them to break down barriers as each party works together on a document. This could also cut down on all of that time spent traveling to meetings."

"Great idea, Tanya," I said. "The need to communicate and meet with one another will increase as the world becomes more complex and the pace of change accelerates. This is a great card to ease the burden.

"The key is to remember one of our rules: *Network with all*. Government, like business and education, doesn't need more independent islands of incompatible information floating around. A system for the Interior Department and another one for the Agriculture Department—and never the twain shall meet—won't work. Right now, we've got computer systems within the government that can't talk to each other. During the Gulf War, the various branches of the U. S. military used sophisticated technology, including cutting-edge communications systems. The conflict with Iraq pointed out that most of our own systems were mutually incompatible. Air force and navy fighter planes operating in the same airspace couldn't communicate with each other."

"We won the war—that's what counts," Maury said.

"We won the war, but did we learn a lesson? It took a lot of fancy technical footwork to make it happen," I responded. "Here's another nonmilitary example: An IRS auditor who's working on a complicated return must obtain documents from six or more different departments, none of which are connected to a central computer system. The data has to be retyped by hand into the auditor's own computer, and, to get the material in the first place, he or she has to formally request it by writing letters to the appropriate coworker."

"Don't even mention the IRS, it makes my blood boil," Tanya said. "Our tax accountant filed my taxes electronically last year, and, of course, the return went off into outer space, never to be found again."

"I understand that the IRS is working on an $8 billion plan to modernize its computer system," I said. "Maybe they'll have the capability to find your return one of these days. I certainly hope you have better luck than the Pentagon, which has a major new computer system in the works for its military hospitals. Anybody want to bet on whether the IRS and Pentagon computers will be compatible?" All I got was a derisive snort from Glenn, so I went on with my story: "Anyway, they are going full steam ahead despite the fact that an earlier version used at Walter Reed Army Medical Center interfered with the doctors' ability to locate medical records and lab results. One physician got on the system's electronic bulletin board and warned that the military was heading for 'certain disaster,' but the justification for continuing is that no patients have actually been harmed by the system yet."

"I hate to say this, Dan, but some of the problem can be blamed on the mainframe industry," Maury said. "We've got the inside track on a lot of this government business. But with these massive systems, a certain amount of flexibility gets lost along the way. To compensate, the system starts becoming cumbersome and complicated. The answer is to devote more time to training so that the operators can extract the information they need."

"But you've got it reversed," Lonnie complained. "You're asking the customer to fit the product rather than making the product fit the customer."

"I'm afraid we're stuck with that sort of thing in large organizations like the government. The trick is to get everyone singing off the same sheet of music," Maury said.

"Isn't it true, though," Lonnie asked, "that the sheet of music is often selected by the information systems department people who understand computers rather than the accountants or administrators who will actually be using the mainframe?"

"The people who write the check for the system get to select its configuration," Maury replied. "Unfortunately, there is usually not enough communication between the information specialists and the end-users. It's not our fault that the programming 'techies' and accountants don't communicate well with each other."

"It's not your fault, but it is your problem when the end-users—like the scientists at the National Institutes of Health—refuse to use IBM's $800 million state-of-the-art mainframe in favor of their user-friendly desktop PCs."

"Actually that's IBM's problem, not mine," Maury said. He picked up his pen and doodled a few lines on a notepad. "It wouldn't be a bad idea to require officials who are making these important procurement decisions to attend seminars like this. Probably every government agency should have an office—maybe just one person—running a technology clearinghouse that keeps everyone posted on the latest developments."

"All right, let's suppose I'm NIH's technology bumblebee," I said. "My goal is to collect the information nectar and cross-pollinate it throughout the organization. I might say to one department, 'Here's some interesting information on neural networks that you might be able to use.' Or, 'If there's a problem with using a mainframe, why not consider a superserver?'"

"You would have to mention superservers," Maury said.

"I take it superservers have nothing to do with McDonald's?" Lonnie asked.

Maury picked up on the question before I could answer. "Unfortunately not. Everyone agrees that mainframes are more reliable than PCs. But the ability to tie PCs into local area networks (LAN) has allowed the low-cost mimicking of mainframe capabilities. These superservers have come along, using cheap PC technology to run large programs and juggle files on bigger and bigger LANs. Now you can hook up hundreds of PCs and printers on the same network." Maury finished his explanation and frowned. "Frankly, I wouldn't trust a huge PC-based LAN with important work. It's like trying to run the electrical system of a Boeing 757 by jumping it off the battery of my wife's old Vega."

"A lot of users are taking their chances, Maury," I said. "It's estimated that by 1995, the number of PCs connected to LANs will increase sixfold, to 55 million."

"Tell me about it. It won't be long before LANs, PCs, and superservers will be grabbing close to $2 billion a year away from main-frames," Maury admitted.

"Sounds like it's time to serve up some superservers," Brenda said.

"I was figuring to do that. Like most of the mainframe manufacturers, we've been shopping to acquire a start-up superserver operation," Maury said, glancing at me. "But now I don't know—as Dan advises, find out what the other guy is doing and do something else."

Ending the Guerrilla War

"In 1991, the federal government spent $20 billion on new computer systems. That's a rather healthy market by anyone's standards," I said. "And that's just the tip of the iceberg. You've probably heard about the Works Project Administration and the Civilian Conservation Corps during the depression; they were massive government programs to put people to work building an infrastructure. The country badly needed dams, roads, bridges, and parks. Helping the government leave the Information Age and enter the Communication Age offers American business similar opportunities.

"Back in the 1950s and '60s, thanks to government leadership, we built an interstate highway system that fully integrated with each state's highways and roads. Can you imagine what driving across this country would be like if each road was built by a different company adhering to their own proprietary specifications? That's exactly what has happened to this country's *infostructure*—our national, state, and local digital information system. Today's regulations fit an America that no longer exists. Technology alters reality, but have policymakers noticed? I don't think so. Many have entered public service from the field of law. This used to be good. Think for a minute, though. How is a lawyer trained to solve problems that impact the future? He or she looks for a past precedent. That technique no longer works. Well-intended regulations, written without specific national long-range goals based on our new reality, are undermining the country's economic strength. I believe that's why so many of our best high-tech companies like Motorola have to go to foreign countries to find growth and profit."

I threw a card on the table.

Don't
Fix the
Blame,
Fix the
Problem

"It's time to play a winning hand together," I said. "First, however, we're going to have to get over this hang-up that the government is the enemy. The government is our customer. In turn, we are the government's customers. If we develop mutually supportive relationships, all of us will succeed. The reality of it is that there is no such thing as 'both sides'; we are all on the same side.

"A government that doesn't understand the new tools and rules of the game puts us at a distinct competitive disadvantage, particularly when our competition is supported by well-run, forward-thinking government entities. Japan and Germany are two perfect examples. In both cases, industry and government are not at each other's throats. There is long-term planning, comprehensive market analysis, and resource and credit allocation. Yet on any given day in Washington, DC, the halls of government are swarming with lobbyists, focused on a narrow interest, playing with half a deck, out to create loopholes or write exemptions into legislation for their companies or industries. A few of them come out on top, but it may have nothing to do with a national vision of what the United States needs to be in the year 2000. Why? There isn't one. Instead we get a hit-and-run guerrilla war against the government.

"It's time for government, business, and education to shift from crisis management to opportunity management. We need to form a new relationship based on a common objective. We need to reindustrialize as we reinformationalize for the twenty-first century. The government has always made a good customer. Ask the defense industry. Their engineers and scientists now need a new focus. I think we already have one, but no one is leading the charge."

"I fully agree that government has always made a good customer and the defense industry hasn't been the only beneficiary," Maury said. "Maybe crisis will force leadership."

"Maybe so, Maury," Samantha replied, "but I'd feel a lot better if leadership grew out of opportunity management."

Customers and Constituents

"Entrepreneurial government is the government of the twenty-first century," I said. "Not only selling lottery tickets at the neighborhood convenience stores and other services to make a buck, but responding to the needs of its customer-constituents in flexible and innovative ways."

"Dan, where I have problems with that is the age-old problem of letting government go head-to-head with private enterprise. Inevitably, government ends up using regulatory power, the ability to tax, and the right to print money to create a monopoly for itself," Maury said.

"But that's not the historical experience here in the United States," Doug responded. "The post office was a genuine government monopoly, and the Feds chose not to protect it from private competition aside from holding on to the first-class mail service. The United States Postal Service is getting its brains beat in today as a result, and I think it deserves to. Wait for someone to come along and put a month's worth of junk mail on CD-ROM—multimedia *and* CD-ROM— that will really give them a headache."

"Maybe the postal service should kill its own cash cow first," Tanya suggested.

"The phone companies may do it for them," Maury suggested. "Did you know that you can send a fax late at night cheaper than mailing a letter, what with the off-hour discounts the phone companies offer? Just build a fax machine into all new home phone systems and give a fax to customers with existing telephone service. This would increase calls and revenue and steal the cash cow from the post office.

"My point is that once government gets on a roll, it's literally 'bigger than the both of us'—bigger than all of us," Maury insisted.

"Show me where that's happened?" Doug demanded. "The U.S. government has given away its divine right to monopolies in oil and gas production, timber and other national resources, telecommunications, even banking. We don't even have a federally owned national bank in this country."

"I'll give you one monopoly," Maury said. "Public schools."

Doug laughed. "Some monopoly. We've got the city's most affluent neighborhood in our school district, and 98 percent of those children go to private schools. If the government were the monopolistic monster that it's made out to be, those private schools wouldn't be allowed to skim the cream off the top of the social strata and leave the rest of us to go it alone."

Re-inventing Government

"Let's not get into a debate over private versus public education," I suggested. "A lot of the rhetoric about the dangers of big government doesn't square with current reality. Maybe it did in the past, but we are going to have to rethink our almost knee-jerk antagonism toward government to determine whether the attitude suits a day gone by rather than the future. Perhaps with the new rules and new tools we can re-invent a new user-friendly government instead of the tax-and-spend, red-tape-loving behemoth that we all agree needs reform.

"And will that be good for business? You bet it will, if we can refocus the government's energy, talent, and resources on making U.S. business competitive in the world. We all work for USA, Inc.—all of us. But Washington seems disconnected—out of the loop. Instead of cutting the senior bureaucrats' pay by 20 percent, as President Bush recommended back in the 1992 campaign, give those officials raises based on the improved performance of the industries they regulate. What's the rule?" I asked.

"You get the behavior you reward," Maury replied.

"Close enough," Brenda said.

"But what about agencies that handle health and safety regulations?" Lonnie asked. "You can't have officials at the Environmental Protection Agency trying to earn a bonus by encouraging pollution so that they can clean it up."

"No, they earn their bonus by devising ways and means of reducing pollution and improving corporate profitability at the same time."

"Is that realistic?" Glenn asked.

"Here's an example from your industry, Glenn," I said. "In the early 1970s food prices were soaring out of sight. You remember . . . ?"

"I was still a teenager, but do I ever remember! When lettuce went over fifty cents a head, Dad thought consumers were going to start lynching grocery store managers."

"The National Commission on Productivity was asked to see if it could come up with ideas for restraining costs, which were being pumped up, in part, by shipping bottlenecks. It could take two weeks to get produce from western farms to eastern markets," I said. "What the commission did was to rediscover a technique that had been used in the 1860s by the railroads to ship freight. They called it LIFO—last in, first off. The commission convinced the growers that to meet tight shipping schedules, they would have to pick their crops at times that were inconvenient and nontraditional; in turn, the railroads were persuaded to stop organizing their trains in such a haphazard fashion by using LIFO so that only the last car needed to be unhooked at the next stop down the line. Everybody won. Shipping times were slashed in half; fresher produce got to the consumer; costs were lowered; and the railroads' productivity improved."

I went to the flip chart. "Do you recall that I promised to come back to the word 'orchestrate'? Well, I'm back."

Four Keys to Success

Integration
Flexibility
Communications
Orchestration

"These are the four spark plugs that drive the engine of success for business and for government. What the National Commission on Productivity did was to provide all four ingredients to solve the food-shipping problem. It got all the various parties talking to each other about the problem; common goals and a methodology were worked out; compromises were cut; and a deal was pulled together and made to work—integration, flexibility, communications, and orchestration.

"Where's the commission today? Out of business. We get the behavior we reward. Today we have lack of integration, lack of flexibility, lack of communication, and lack of orchestration. The commission and its officials should have been rewarded for the work that was done."

"Dan, wasn't the White House Competitiveness Council, which was chaired by Vice-President Quayle, doing similar work?" Samantha asked. "It certainly drew a lot of flak before Bush and Quayle were defeated for re-election."

"Yes and no. Yes, in that it was a forum to air disputes over government regulations; no, in that it was a backdoor method to grant exemptions to individual companies that complained the loudest. It was a hit-or-miss approach that lacked *integration, flexibility, communication,* and *orchestration.*

"A better way to go would be an agency modeled after Japan's Ministry of International Trade and Industry. MITI has played the role of orchestrator for Japanese business, and without an orchestrator you can't have everyone singing off the same sheet of music—to use Maury's phrase. The image that I see is the western sheriff's posse that saddles up and rides off in every direction. That's what happens most of the time.

"The Ministry of International Trade and Industry, however, organized the posse during Japan's painful rebuilding process after World War II and on into the present. What's MITI up to now? For one thing, it's looking down the road and seeing a rapidly expanding elderly population. The ministry wants to hook those elderly citizens into a vast digital information network that will provide them with the services they are going to need in their declining years. Instead of looking on this project as just another government welfare boondoggle, MITI sees it as an opportunity for Japanese business. The network is going to include hospitals, insurers, beauticians, fashion designers, telecommunications equipment manufacturers, and other enterprises that can provide services ranging from health care to choosing a vacation spot."

"What's with us?" Brenda asked in exasperation. "We've got a huge elderly population on the way too. The baby boomers are already starting to turn gray; in a little over a decade they'll begin edging into retirement. They don't even want to think about that, so few do."

"Forty years ago, the ratio of active and productive workers to retirees in the United States was seventeen to one," I said. "By the end of this decade the ratio is going to be three to one. In ten short years, 45 percent of all new workers will be Hispanics and African-Americans—two groups that, unfortunately, are showing the largest increases in illiteracy and school dropout rates as well as declining enrollments in college. That's a human tragedy and a national economic calamity in the making!"

"Maybe we should hire MITI to do some orchestrating for us, Dan," Lonnie suggested.

"Not a bad idea," I said. "Or we could put our most successful orchestrators on the job—the U.S. military. They've got quite a track record at working with industry to come up with products that are unmatched by the competition. Since 1960, a full one-third of all research and development funds have been defense related. That's one of the reasons why we won the Cold War. As an economic bonus, military R&D helped launch the electronics, computer, and aerospace industries."

"Well, that's history," Maury said.

"I hope not. Industries with the highest record of research, development, and capital investment have been shown to be the driving force behind growth in jobs and the standard of living," I pointed out. "If we stop priming the R&D pump, we are in big trouble."

"But we're in trouble anyway, and according to your figures we've been investing heavily in defense-related R&D for thirty years. The pump doesn't seem to be working as well as it once did," Tanya said.

"Let me draw you an R&D flowchart," I suggested.

$$$$$	*Yields*	Research	*Yields*	Ideas

"Dollars go in one end of the pipeline and breakthrough ideas come out the other end. Pure research. The system has made us the envy of the world when it comes to producing great ideas. We're good at it. What we aren't good at is developing applications around those ideas to make them into salable products and industrial processes. Aside from building smart bombs, we've left that job to our foreign competition. The United States has been content to work with the old ideas and marginalize the new stuff. It's odd, really. Only 70 percent of nondefense R&D goes into product research. The Japanese spend the same proportion on what is known as process research. They then turn around and use their breakthrough ideas in the process area combined with our breakthroughs in products—and clean our clocks."

"Clearly, we have to fine-tune R&D," Glenn said.

"I like my word better—orchestrate. That's what the government can do best, just as the military demonstrated during the Cold War. The object is to create a demand for research, not just a supply. Let's create a research network that links up universities, corporations, and government laboratory facilities, such as Los Alamos and Lawrence Livermore. The government should encourage collaboration that crosses competitive boundaries, use tax incentives to stimulate innovation rather than emigration, and seed a wide variety of research—big and small—that impacts on every key sector of the economy."

"Don't forget science, engineering, and math education," Doug suggested.

"How could I, especially after I read about a study in *Business Week* magazine that gave me an idea for this chart."

More Lawyers or Engineers?

- Increase Engineering School Enrollment by 50%
  Yield = 0.5% Increase in Economic Growth
  Note: Would require a 10% shift in the current student population.
- Increase Law School Enrollment by 50%
  Yield = 0.3% Decrease in Economic Growth

Successes and Failures

"We need a plan," Lonnie declared. "How about no low-tech businesses in the United States by the year 2005?"

"How do you define high-tech versus low-tech?" Brenda asked.

Samantha had the answer to that one. "Simple. If they're not playing with the new tools of technology and the new rules, they're low-tech."

"Okay—now implement the plan," Maury said. "Give me step one."

Everyone around the table turned my way. "Why look at me? You're doing fine."

There was silence. Finally, Samantha said, "The first step is the toughest."

"I'll help with a couple of new rule cards."

"Where did you go to college, Tanya?" I asked.

"University of Illinois."

"What's the difference between the University of Illinois and Harvard?"

"About $15,000 a year in tuition."

"Anything else? And I don't mean ivy."

"Illinois is a land-grant college, like a lot of other state universities that received money and land from the federal government back in the nineteenth century." As soon as Tanya answered she lit up. "That's the success of the past! The agriculture extension service grew out of the land-grant college system. Research facilities were set up on college campuses to help farmers cope with changing conditions after the Civil War. We should establish a high-tech extension service."

"It's already happening in twenty-three states, which are spending about $50 million a year," I said. "But the Japanese are doing the same thing and investing $500 million. Here again, they're making the most of an American idea. When the agriculture extension service was set up in 1914, American agriculture was lagging behind Europe in terms of the methods and technology used on the farm. There was a dramatic increase in productivity after the government went to work introducing farmers to the latest techniques and hardware. There are 380,000 small businesses in the United States, many of which are in the Dark Ages when it comes to technology. Just as those farmers were in 1914. I've seen figures that indicate that we'd have to spend about $480 million to launch high-tech extension services in all fifty states; at this very moment the budget for the agriculture extension service—and this is not an agricultural economy anymore—is $1.2 billion."

Forming Partnerships

"As for re-inventing the failures of the past, I'd re-invent the cartel. And isn't that a shocking idea! Cartels, in the days of the robber barons and John D. Rockefeller, were a pretty good idea in that they were created as a mechanism to keep fledgling industrial capitalists from cannibalizing each other. But things went too far, and the cartels disrupted the free marketplace. Teddy Roosevelt and the trustbusters came along; and ever since *cartel* has been a dirty word.

"So, call the new cartels by a different name—'consortiums.' I can think of three of them that could serve as positive role models for dozens more."

United States Consortiums

1) Microelectronics and Computer Technology Corporation
   15 U.S. computer companies involved in developing electronic packaging, software, and parallel processing architecture.

2) The Open Software Foundation
   Computer hardware and software companies working on standard interfaces.

3) Sematech
   A consortium of U.S. semiconductor manufacturers founded by matching grants of $100 million from each of its members and the federal government.

I said, "The government will need to change its attitudes as well as its laws to allow for more of these collaborative enterprises to develop. This isn't 1900. We are no longer a nation brimming with natural resources and cheap labor. That was the era of cast-iron capitalism. Today we are engaged in intellectual capitalism. We'll still make things. But they'll be made smart. If not . . . " I paused and laid down a card.

"Lonnie will replace the bald eagle as our national symbol."

12

Whole Life

Winning the Next Hand
Before It's Played

"Samantha, what do you sell?" I asked, after dealing cards to each of the players; it was the first hand of the evening. I knew the answer to the question, but I wanted to shift everyone's attention to the group's youngest member.

She gave me a strange look. "Insurance. Life insurance in various packages, shapes, and sizes."

"No, what do you really sell?"

Samantha was young, but she was a well-trained, intelligent salesperson. She gave a classic response. "Myself."

"You're getting closer to what I'm looking for, but you're still not there. Anybody else want to take a stab at it?"

Brenda gave it a try. "Dan, when we first started these games you said that Samantha had a tough job because she was selling money for future use . . . is that it?"

"Gee, Brenda, you've got a better memory than I do," Lonnie said. "That one went right by me."

"Maybe in that case we should pause for a little refresher course," I said. "Samantha's customers assume that tomorrow is going to be just like today—even though they know better. Old age, illness, and death are out there waiting for all of us. Even so, it's easier not to think about it. When Samantha comes along, they're not particularly eager to take money out of their pockets today for use tomorrow. What she has to do is convince them that it's necessary.

"Back to my question—what is Samantha selling?"

"She's selling the future," Maury said.

"That's definitely part of it," I replied. "But first, before she can sell the future, she has to sell trust. Insurance agents would never write a single policy without first selling *T-R-U-S-T*." I spelled out the word.

"Think about the challenge. The rest of us ask the customer to pay for a tangible product: a computer, a hospital bed, a case of dog food, a rosebush, a gallon of milk. They hand over the money, and we give them the product. Samantha's customers write out a check and may never see the real product. Many of them will be dead before she actually delivers."

"Dan, that's one of the reasons I like to sell whole and universal life; I emphasize the savings and investment features. The customer can get a life insurance policy and an investment vehicle for retirement that offers great tax benefits. The new insurance products are not like selling cemetery plots."

"But really, now, isn't insurance just another form of gambling?" Glenn asked. "I bet that I'm going to live long enough or stay healthy enough to earn more in benefits than I pay out on premiums."

"Even if you look at it that way, Glenn," I replied, "the gambler must still trust his or her bookie to pay off. What's the point of betting otherwise?

"One of the reasons that social security is under such pressure these days is that an increasing number of people doubt that the federal pension program will still be solvent when they retire," I added. "Politicians and government officials are not effective when it comes to selling trust. They're not believable. One reason for this lack of believability is that they don't have a vision of the future that is powerful enough to counter the fear that social security taxes are being dumped into a black hole. Remember what we talked about last time? Many government officials are lawyers, and lawyers are trained to look to the past to find solutions for problems that impact the future. In short, they tend to foster crisis management. When planning for the future, remember that the present is obsolete."

I went to the flip chart. "The Walt Disney organization is full of opportunity managers. Here's one of Disney's four guiding principles."

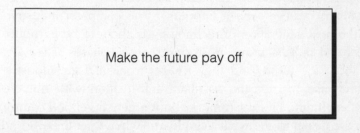

Make the future pay off

I asked, "Isn't that what an effective insurance agent does? Isn't that the implicit promise that he or she makes? And isn't that the reason the pessimist says, 'What's the use? I'm not going to bother investing, buying an insurance policy, saving for my old age. The future is too uncertain, so I'll worry about today now, and tomorrow later.'"

"So, when you say that I'm selling trust," Samantha said, "I'm telling a customer—'Trust me, I'll work with you to make the future pay off for us both.'"

Tanya raised her hand. "Isn't that what we all should be doing?"

Lonnie reached for the deck of rule cards. "We've used this one before."

Sell
the Future
Benefit
of What
You Do

"Lonnie has found precisely the right card," I said. "If Samantha can't sell the future benefit of her insurance products, she's out of business. We all are. During times of rapid change, a successful business relationship depends on the vendor's guarantee that a product or service will be of future use; in that way, insurance is the paradigm for all other industries. Sales resistance grows in a direct relationship to the increasing pace of change and to growing doubts that the product, or its parent organization, can keep abreast.

"Maury, you've seen it happen in the computer industry. Whenever the next generation of technology is about to be introduced, customers will pull back and postpone their purchases. They are uncertain about the future and they freeze. Ironically, the industry makes things worse by refusing to share information with the customers about the future. I'll hear rumors about a new notebook computer and go to the dealer to see what I can learn. When I get there, I find someone who knows even less than I do."

"Dan, that's standard practice. We don't want to sabotage the existing product line while there is still life left in it," Maury explained.

"And that undercuts the credibility of your own dealers and sales force by sending me the message that the product currently on the shelf, one that might fit my needs perfectly well, might not really have value to me in the future. The dealer who shrugs his shoulders when I ask about next month's product launch is totally useless to me. Without someone to act as my personal futurist—someone like Samantha who will say, 'Trust me to make the future pay off for us both'—I'm going to assume the worst."

Obeying the Law (of Averages)

"Dan, you said Disney had four guiding principles. What are the other three?" Glenn asked.

I went to the flip chart and wrote:

Guiding Principles of Disney

1) Make Tomorrow Pay Off
2) Free the Imagination
3) Build with Lasting Quality in Mind
4) Have Fun

"Let's focus on number three, 'Build with lasting quality in mind.' Doesn't that sound contradictory? How do you build with lasting quality if this rule applies?" I tossed out a rule card.

"By redefining quality," Brenda said.

"Good! But who does the redefining?" I asked the question and waited for a response.

"Evolving technology," Samantha said.

"On the surface that might appear to be true," I said. "Back in the 1980s, Xerox decided to invest in a top-notch service operation, the best in the industry. The Japanese, Canon in this case, knew they would have trouble competing with a company as good as Xerox, so they didn't. They went beyond the competition by using technology to change the rules of the game. Canon introduced copiers that *didn't need service*. Essentially, they identified the parts that could break down and put them inside a toner cartridge. The customer was already used to replacing toner; now, without them even knowing it, they were also self-servicing the machine. Each time the toner was refilled, the most vulnerable parts were also being replaced. Thus, both service and quality were redefined. The competitive focus was shifted from a product supported by a terrific service department to a copier that didn't need a terrific (and expensive) service department.

"Now, I'm just guessing, but I'll bet somebody at Canon went out and asked customers 'What do you dislike most about copiers?' And the answer amounted to 'Waiting around for the service technician to come and fix the stupid thing.' The next step was to use his or her imagination to ask 'What is the ideal copier in the eyes of the customer?' The rest is history. In effect, identifying and fixing current problems—problems that were regarded as inevitable—allowed them to redefine the product. Xerox combined traditional competitive practices such as pricing specials, enhanced durability, quality, and beauty—techniques its competitors were also using—and dispensed with the need for a service technician to jump far beyond the competition.

"Don't forget the power of this rule card, because whenever you put it into play, you can change reality in the eyes of the customer." I threw out a card.

"For generations, American business school graduates went out onto the job convinced that quality was equivalent to high cost. The reason for that assumption was this rule, and it is one that still applies."

I continued, "The more time you spend in physical contact with the product or holding the customer's hand, the more it increases costs, both for you and, in most cases, for your customer. The traditional linkage between cost and quality comes either from the need to inspect for flaws in the manufacturing cycle and to correct them, or from customer returns and complaints. Thus, layers of redundant systems are imposed to filter out defects after the fact. And you know what? Ultimately, it is almost impossible to produce a perfect product on a consistent basis. There are enough variables in the manufacturing process, or the service industry for that matter, that the law of averages dictates that the pattern will not remain constant forever.

"What's the answer? I am afraid that for most manufacturers in the United States the answer for far too many years has been high cost and a tolerance for low quality. This used to work. The customer was subliminally conditioned to expect problems. Life isn't perfect, after all. Bringing your new car back to the dealer several times within the first few months of the purchase became standard practice. We could get away with it as long as everyone was playing the same game or not playing at all.

"Enter the Japanese.

"The historic breakthrough was the realization that by completely redefining the design process rather than correcting defects in the finished product after the fact, they could lower overall production costs, reduce customer returns, and improve quality. This led to the Japanese insistence on *constantly* improving the process—not just occasionally."

"What did they do?" Doug asked.

"They played one of the new cards," I replied. I put it on the table.

"Whether they are building a new manufacturing facility or preparing to produce a new product, they link design engineers together with suppliers, distributors, salespeople, and every other aspect of the entire process. The U.S. business response was typical: to imitate instead of innovate. We will work hard at copying the Japanese, but in the meantime the competition has already moved ahead to implement the next round of innovations. The imitators, as a result, always end up lagging behind the innovators.

"It is now quite evident that an improved process will lead to dramatic changes in the product. Now, *that* could be a problem. Can anybody tell me why?"

"The process takes over and dictates what kind of product will be sold, whether or not there's a market," Lonnie replied.

"Exactly right," I said. "Henry Ford demonstrated this when he allowed his assembly line to dictate the color of the automobiles he was offering the public. He said the customers could have any color they wanted as long as it was black. It gave Chevrolet and other competing carmakers an opportunity to squeeze into the market by offering the public a choice of colors.

"Take Swiss watches as another example. No, this isn't the highly publicized story of how the Swiss watchmakers developed the digital watch and didn't do anything with it because they failed to see the market value. The digital part of the story doesn't reveal the strategy. The Japanese knew they would have a tough time competing with the Swiss for the global watch market, so they didn't. They knew that the Swiss manufacturing process had evolved over decades, insuring the production of extremely accurate watches. The Swiss took great pride in how they defined what a good watch should be. Their pride was actually a weakness, because it would keep them from stealing their own cash cow. The Japanese knew that in order to go beyond the competition, they needed to change the rules of the game by redefining the function of a wristwatch. They perceived that the customer was ready for a watch that did more than just keep time. By holding down the cost of the watch, the customer could afford to have several. And by creating a wide variety of highly specialized watches, the customer would want more than one. The Swiss were so committed to their traditional process that they just kept turning out extremely accurate, high-quality watches, and the market all but vanished.

"The Japanese embraced change in the manufacturing process as a way to improve product quality and as a way to reduce costs. By making change their best friend, they developed the flexibility to meet the customer's constantly changing requirements. Consequently, they can redefine quality in the eyes of the customer again and again."

I pointed to the flip chart. "Go back to Disney: 'Build with lasting quality in mind.' It's lasting because there is a constant evolution to keep abreast of the customer's requirements, not the requirements of the Disney organization. Nothing ever freezes in place. The strategy recognizes that during times of rapid technological change—the times we're living in today—the customer and the competition will always be constantly defining and redefining quality.

"Therefore, a manufacturing or service industry is not driven by a particular product or a process. It's driven by the changing needs of the customer—and building change into the process and into the product is therefore essential. This is why I have been hammering since day one of our games at the rule—*Make change your best friend*.

"As for Disney's guiding principles number two and number four—'Free the imagination' and 'Have fun'—we need to once again consider the following rule; it should be an old friend by now."

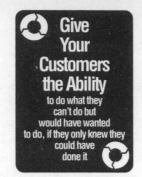

Give
Your
Customers
the Ability
to do what they
can't do but
would have wanted
to do, if they only knew they
could have
done it

"Most customers know what they want today because they base the decision on what they wanted or needed yesterday. To avoid being trapped in the past, Samantha and the rest of us have to accept that change is occurring and go all out to anticipate the effect of those changes. Her success as an insurance agent is directly dependent on her ability to anticipate and fulfill the future needs of her customers. Those needs are going to be there, whether or not the customers know it, and whether or not Samantha positions herself accordingly."

"Dan, I see why you've been juggling three sets of concepts throughout these games," Doug said. "If we just bought into the need to understand new technology, it wouldn't be enough. And even latching on to some of the new rules doesn't do it. Without using change to our advantage—and welcoming it—the whole thing falls flat."

"Go to the head of the class," I said. "The linkage is this: *change in conjunction with technology*. Since I run a technology research firm, it may seem that I'm interested in technology for its own sake. Although I do have a natural passion for science, my deepest interest is in its creative application. What fires me up is the phenomenon that we label with a simple and deceptive word: change. The ability to adapt to change, as Darwin recognized, determines the success or failure of a species. I've always wanted to know where change was coming from. What drives it? How and why one species adapts and another perishes.

"Within the animal kingdom, humans are unique as a species in that we can apply technology to create permanent changes that are capable of altering reality itself. Change is mediated through technology, from the most primitive form to the most sophisticated."

"My cat mediates the changing requirements of her appetite through the technology of the electric can opener," Lonnie said.

"There are a few animals that have gotten the hang of some basic technologies," I said.

"And a long list that haven't," Tanya added.

"That may be the bottom line," I said. "Survival for any species depends on the ability to come to terms with the prevailing change agent, whether it's a drought that destroys crops or a chain saw that cuts down the rain forest. The most powerful change agent we have is technology. Descartes said, '*Cogito, ergo sum;* I think, therefore I am.'

"Try this: I think—I change—therefore I am."

Change and Aging

"Disney's guiding principle number four, 'Make it fun,' confirms what I've always thought about change, Dan," Samantha said. "Change is fun. I go on vacation because I need a change of scene. I try a new sport for both the change and the challenge. Working with different customers every day keeps me sharp. I wouldn't want the same old routine."

"I felt the same way at your age," Maury said.

"But is fear of change a function of age?" Brenda asked.

"No, there are many young people who fear change. But traditionally the ability to change depended to a great extent on flexibility and mobility. If 500 years ago your only well dried up—and that was quite a change—you might have to pull up stakes and move hundreds of miles. An old person's ability to do that was limited. During the Industrial Revolution, physical stamina was also important. As we get older, we become less flexible both physically and mentally. Our routines become more important to us, and we find comfort in knowing what to expect. We don't want surprises. In both the distant past and more recently, the elderly have been at a disadvantage. Therefore, change was something to fear, especially once individuals passed their more youthful, physically vigorous years.

"That primitive instinct still echoes within us, but technology has compensated for lack of flexibility and mobility or declining physical strength. The elderly need not fear change. In our last game, I was talking about Japan's Ministry of International Trade and Technology and how it's working on a digital information network for the elderly. In the 1980s, nursing home patients in the United States were given a demonstration of what a personal robot would be able to do in the near future. They were then asked what they thought of robots. Most wanted a robot! They said: 'My hands have arthritis and it hurts me to open a door. I would love for a robot to do those types of things.' Some said: 'Since robots can be programmed with a personality, I would have one to play games with.' These are perfect examples of how technology could work to our advantage no matter how old we are."

"Maybe I'm going to start sounding a little spacey," Maury said, "but if you track out the dynamics of change as you were describing them, Dan, if we can change the way the customer thinks, and the customer can change the way we think, and that changes the process, which in turn changes the product—and on and on and on—isn't change almost the fountain of youth?" Maury chewed on his lower lip after asking the question, as though he regretted having brought it up.

"We've been talking about re-inventing products and re-inventing corporations. Change also allows us to re-invent ourselves. If I see myself as a work in progress until the day I die, maybe that amounts to re-inventing the fountain of youth.

"I've referred to retirement as shifting to a time of re-engagement. In the future, re-engagement won't necessarily occur at sixty-five years of age. It will happen at twenty or thirty or forty. There will be several re-engagements over the course of a lifetime. And Samantha is in a profession that will play a central role in that process of re-engagement. She mentioned selling cemetery plots a little while ago. That's what insurance used to be. It was death and disaster oriented. The customer got old, retired, disengaged, and died. But if there is going to be re-engagement, Samantha has to change the way her customers think about their retirement years and how it relates to life insurance."

"Nevertheless, Dan, death or disability payments are still the object of insurance," Doug said. "How do you get away from the Grim Reaper?"

"Let's ask Samantha," I suggested. "When you sell a new life insurance policy, what's your underlying vision of the transaction? Are you sitting at your desk looking at a dead man or a dead woman? Is the customer wearing a set of wings and a halo?"

"That's never happened. I usually picture an unbroken stream of premium checks flowing into our bank account—which sounds pretty mercenary. But keep in mind that many customers are also using this money as a retirement savings plan. I care about my customers, and if I have any thoughts along the funeral line, it's usually a picture of a kid in cap and gown graduating from college even though Mom or Dad isn't around to see it happen. Dan's right, I'm selling the future."

"If that's your product, you better know a lot about it if you want your customers to *trust* you with their future. Where do you start?" I asked.

"Where everybody else started, with a plan."

"Good. Because Samantha is the last one to play, I want to tell you one of my favorite planning stories. A study was done of the Yale class of 1953. Twenty years after graduation, in 1973, the researchers evaluated the level of success in the class. They found that 3 percent of the class had 97 percent of the net worth of the entire class. In addition, the researchers found that those 3 percent were happier, healthier, and had a better family life. The researchers could only find one difference between the two groups. All of the people in that 3 percent group left college with a *written* set of specific goals and aspirations."

"Okay, I didn't go to Yale, but here's my plan," Samantha said, eagerly looking at her written goals, rubbing her hands together. "I want to continue to be a member of the insurance industry's Million Dollar Round Table, every year, for the rest of my career. That means I have to write more than $1 million in new policies each year to qualify as one of the top 5 percent of all agents in the world. And I'm going to sit at the Top of the Table, which is reserved for the top 2 percent in terms of total policies sold." She paused. "But there's one problem: I haven't the faintest idea of how I'm going to do it."

For Form's Sake

"I'll give you a choice of two cards, Samantha. I made them octagon in shape so you wouldn't confuse them with the rule and tool cards. One of them is the *traditional way* to earn new business," I said.

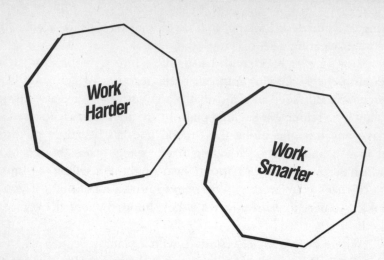

"There's no choice," she said. "I'll take the second one."

"Fine. Now make a pair with another rule card that we've already used."

"Well, I didn't choose 'work harder' because if anything, I'd like to work less. I'm already putting in a sixty-hour week."

Lonnie, Brenda, and Glenn all reached for the rule deck simultaneously. Lonnie got it first. "May I?" he asked.

"Deal."

"Now form a pair with a tool card," I said.

"Easy." She reached for the tool deck.

Electronic Notepads

"I can't use a laptop or a notebook computer because the typing would break the rapport I'm trying to build with the customer. They don't mind when I use a pen and paper. They are accustomed to that, so there is no distraction," Samantha said. "We are selling trust, and I've got to have a maximum amount of face to face time without anything interfering."

"How about pen-based input?" Maury asked. "An electronic notepad with electronic pen input would be perfect."

"If it would speed up the process and not be disruptive—great. We could fill out forms and questionnaires using a method the customer is already comfortable with. It would be neat if I could access the computer for a particular form instead of having to push a bunch of buttons to find it. And not only would I like to install my forms in the notepad, but I'd want to be able to use it to call up brochures and other chunks of information that are normally paper-based. I feel like a packhorse lugging that stuff around, and usually what the customer wants to see is back in the office anyway. I could use a portable miniprinter to do a printout."

"I think Samantha should also install a digital cellular phone connection to link the notepad to her insurance company's database," I said. "That would probably speed things up when it came to qualifying customers. A lot of the paperwork could be executed on the spot."

"I dread it when my insurance agent visits," Tanya said. "He's always digging in his briefcase for paper, and when I ask him for information he doesn't have with him, I have to schedule another appointment."

"You're onto a major industrywide problem," Samantha said. "If we don't make the sale while we are with you face-to-face, it is likely that the sale will never happen. And it's taking longer to write and process each policy. Time is money; excessive time cuts into the profit margin. It's a real dilemma. Dan's right that 'High touch means high cost.'"

"If I only had one face-to-face meeting with my insurance agent, it would be all right by me," Maury said. Glenn agreed, and so did I.

"It's this card," I said.

Time
Is the
Currency
of the
'90s

"Some of your customers may need the personal contact that comes from face-to-face meetings. What I need, and I suspect what Maury and Glenn need, is time. Sell us time as well as insurance. It's something your competition isn't doing. They all assume that today is like yesterday, and we've got hours to burn."

"But how do I know which customer is time-sensitive?"

"Ask 'em," Maury replied.

"Or, during the course of your first face-to-face meeting, drop some opening lines like, 'I seem to have more to do and less time to do it. Have you noticed this, too?' You might learn a lot," I said. "Personally, I've found that meetings with my insurance agent are so constrained by time that I come away with very little new information about products and services that I might otherwise be interested in."

"What if I found out which of my customers work with a PC and let them link up to my office's advanced expert system?" Samantha asked, tossing the card out on the table.

Advanced Expert Systems

"I don't have one now, but a system that includes all the products I handle could save many of my customers time and give them a new advanced service. I could probably get one put together for a relatively low cost by working with students who are majoring in this subject. Then I could sell it to other insurance agents to cover the investment. The questions and answers that I typically deal with in a face-to-face meeting could be handled by the system."

"I tell you what, send me that system and a software-based calendar program instead of one of those wall calendars, and you've got my business," Glenn said.

I tossed out a tool card.

Object-Oriented Programming

"You could use object-oriented programming to customize the software or, once again, hire a student to do it," I said.

"Design it to remind users to call you for an annual consultation on their birthdays," Brenda suggested. "Or, if their child is graduating from college, it could ask them to open an icon that would provide some information about life insurance for young people."

"I like that, Brenda," I said. "It reminds me to show you this rule card."

I continued, "Build change into the software program. The latest information on a young person's life insurance policy might have changed since the calendar program was devised. Have the program flash your fax number—an interactive fax number. The customer calls and is voice-prompted to push the numbers on the keypad of the Touch-Tone phone: 'If you are calling about universal life coverage, push one . . . individual coverage for family members, push two.' He or she makes selections, and the appropriate information is automatically faxed. There's no need to talk to anyone or take time out for a meeting with the insurance agent."

Samantha reached for another card.

"This will help me leverage time. I can do business from every-where and anywhere, including my boat on the weekends, as long as I know that I can get a static-free connection. A lot of my customers think that I work twenty-four hours a day anyway."

Her next card was a natural.

"Imagine a paperless insurance office! By converting my paper-based information to digital information, I can have better access to it. I'll bet there's a lot of knowledge buried in all my files. I should be able to get it quickly, anytime I might need it. That alone will help me cut cost and hassle," Samantha said.

"That pairs off with the rule card I played earlier tonight: *High touch means high cost,*" I said. "The more you have to touch the prod-uct by filling out paperwork, meeting with the customer, dealing with the insurer's administrative staff, the higher the cost."

"Wouldn't this card help in that respect?" Lonnie asked.

"But most of my customers aren't going to have multimedia capability at home or in their offices right away," Samantha responded. "They'd have to come to my office, and that's not building a better path to the customer."

"I have an idea," Doug said. "I've read about banks setting up multimedia kiosks in shopping malls to do everything a bank can do, including credit approval for mortgages and dispensing financial advice. Get together with travel agents and split the cost of doing multimedia kiosks at various places around town. The tie-in would be flight insurance for travelers and vacation cancellation insurance. You could find an assistant to write those policies, since it isn't your specialty. Go from there and offer full life coverage with a sound, video, and text presentation."

"And you'd be ready when this card came along," Tanya said.

"If it's good enough for Blockbuster Video, why not life insurance?" she asked. "And then this card."

Samantha said, "In every case, I'd be *building a better path to the customer.* And I like Doug's idea about offering additional services in conjunction with travel agents. I'm glad you didn't suggest working with the banks. I'm afraid they're going to go into competition with us if they can get away with it. What I'd like to do is go into competition with them by taking the advanced expert system card and turning it into a financial planning service as well. I could see hiring a certified financial planner to work in my office as a human backup to the advanced expert system."

"That's the way to go, Samantha," I said. "You're changing your role as an insurance agent. If you are selling money for future use, it's logical for you to move in the direction of financial planning. You're using new tools to change the rules as you define a new role and accomplish ideal goals."

Samantha nodded. "I've also been trying to think of a way to position myself within the insurance business to take advantage of my already established relationships with major insurance companies. When Brenda said she wanted to set up a system at her hospital using bar-code readers that would scan a patient's wrist bracelet to record the drugs and other treatment they receive, I thought there might be a possibility . . ." She let the sentence trail off.

"Help underwrite the cost of the system, and plug in and sell the insurance-related data back to the insurance companies," Maury said. "Connect them into the system through fiber optics or direct broadcast satellite on a real-time basis. Charge them a licensing fee per hospital."

"We'd want a piece of the action, too," Brenda said. "And we could sell Samantha optical disks with pertinent data that she could use to prospect for new business."

"Or to screen out high-risk potential customers," Samantha said.

"Now you're into an invasion-of-privacy area," Lonnie said.

"How about a lowering of insurance rates?" Samantha snapped back. "If we can avoid high risks, we can cut the costs for the majority of policyholders."

"By cherry picking. I don't want you snooping around in my hospital records and raising my premiums, thank you," Lonnie replied.

To keep the debate from getting out of control, I suggested that the more information the insurance company had, the more efficiently they could spread the risk without having to take wild gambles. Lonnie wasn't completely satisfied.

"I think this aspect of technology will end up being abused," he said.

"A statement I made during one of our earlier games is really a new rule," I said.

"I can assure you that, like it or not, when a new technology reaches the application flash point, it will be played. If our country doesn't do it, others will. We can, and we should, give close scrutiny, apply reasonable regulations, and take corrective action when mistakes are made—and mistakes will be made. But there's no way to get the genie back in the bottle."

"If we don't open the bottle . . . "

"The bottle is wide open. It's been smashed open. We all know that there will be abuses of this new technology. Some will use it to destroy; others will use it to build. Technology is not good or evil. The central question is, how will humankind apply it? Things will go wrong. But we can't let that stop us from making positive uses of these important breakthroughs. I remember as a kid visiting a shoe store and putting my feet into an X-ray machine to see if my new shoes fit. That machine zapped me, and probably millions of other people, with a dose of X rays we now know was about 600 times higher than currently accepted safe levels. Was that an evil plot? No, it was a mistake. Should we have banned all X-ray devices? Of course not! Should we have taken steps to make sure we're not zapping our feet in shoe stores? Definitely. And that's what we did."

An Ounce of Prevention

"Since we've lifted the lid on the uses and abuses of technology, I think we should go for broke. What can you do with this card, Samantha?" I asked.

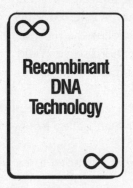

∞

Recombinant DNA Technology

∞

She took one look at it and shrugged. I looked around the table. "Anybody else want to try?" There were no takers.

"We've talked about genetic markers that indicate hereditary predisposition to certain diseases and chronic conditions. Use rDNA to go a step beyond financial planning to another new definition of what constitutes an insurance agent: a life-style counselor."

"A life-style counselor?" Samantha asked.

"In the near future, we will be using a blood test that will screen for genetic markers that relate to the 4,000 known genetic diseases. At first we will test for several hundred diseases. After gene identification research is farther along, we will test for thousands. The goal will be to identify genetic predispositions to inherited diseases. Working with Brenda's hospital, you could establish a program for your life insurance customers who have tested positive for certain inherited diseases, and, based on the results, provide professional advice on measures to prevent the diseases from developing. At the same time, you'd be in a position to lower premium rates by reducing the variables. And, by promoting preventive medicine, you will reduce medical costs by preventing the problem in the first place. You will be better serving your customer's future needs."

"That's the Lonnie scenario, only worse," Glenn said. "You're not even sick, and the insurance company denies you coverage because of an alleged genetic defect."

"The savings from preventive measures alone would give insurers the cushion they need to write the policies and spread the risk," I said. "A genetic predisposition does not guarantee that the disease will ever develop. If the blood test indicates a genetic bias to lung cancer, the person could be 105 years old before it ever appears. However, if he or she smokes, lives in a polluted atmosphere, or works in an industry with airborne emissions, the disease could come on at an early age, costing the insurance company tens of thousands of dollars in health care costs.

"By working with the customer, Samantha would be selling the future—a healthier future—and allowing her industry to make a reasonable profit."

"What about the person who refuses to take the test?" Lonnie asked.

"They'll probably end up paying higher insurance rates because there will be more risk. They might even lose their coverage. Should an insurer be required to issue family coverage to a mother and father who refuse to immunize their children against polio or scarlet fever?"

"On religious grounds?"

"I don't care on what grounds. If I take certain actions that put my health at risk, like smoking when I've been told I have a genetic predisposition to lung cancer, and an insurer offers to work with me— provide a hypnotist, a psychologist, an acupuncturist, whatever—and I say, 'No, I'm going to keep smoking,' the insurer is crazy not to raise rates or cancel my coverage. Under those circumstances, they'll go broke or keep raising the rates on everybody else until we all go broke."

"Dan, I don't see the difference between a blood test for genetic markers and a blood test for high cholesterol," Maury said. "If you refuse to make an effort to bring your cholesterol rate down, why should I gamble that you're going to luck out and avoid a heart attack?"

Doug joined the debate by saying that he was prone to catching colds. "I pick them up from my students all winter long, and several of my teaching colleagues sail straight through in perfect health. If a blood test was developed that could determine that I was genetically susceptible to rhinoviral infections, that would help to explain why I get sick and others don't. With that knowledge, I would be obsessive about staying out of drafts, getting plenty of sleep, and taking vitamin C. I would also like to know if I was genetically predisposed to get something that might actually kill me. With that knowledge, I could alter my life-style and put off dying by a few decades."

Lonnie smiled. "I guess that's what Dan was talking about when he told us: *Solve your customer's future problems today*."

13

Competition and Combination

Two (or More) of a Kind

When the group arrived for our last game, three strangers were waiting in my dining room. I avoided introducing the newcomers until I had made an announcement.

"Tonight is the big casino. We'll find out for sure whether you have learned to play with the new rule cards and the new tool cards. To make it interesting, I'd like everyone to throw $1,000 into the pot." There was a gasp of amazement. Even Maury, who could easily afford a stake of that size, was nonplussed.

"Don't worry; if you don't have the cash I'll take an IOU."

". . . look around for the sucker," Lonnie recited the axiom. "If you don't see one, get up and go home because you're the sucker."

"You're a sucker if you don't know there are new rules and new tools," Maury said. "None of us are suckers." He pulled out a blank check from his billfold, signed it, and tossed it into the center of the table without bothering to fill out the amount. The others used their yellow legal pads to write IOUs. Tanya looked distinctly queasy.

"All right, everybody has anted up. My friend here, will deal," I said, stepping back into the corner to watch as one of the three men who had been waiting quietly came forward. He used three decks, delivering two cards from each. At first, there was a noticeable sigh of relief. I walked behind Glenn and saw that he had been given two familiar rule cards on the first pass:

Glenn looked pleased. The other players were also comfortable with their cards. On the next round, dealt from the second deck, I saw that Doug received cards that would challenge his skill:

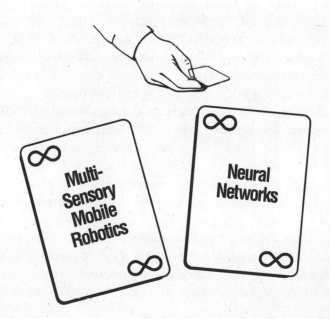

Doug looked like he could handle it, though, and his rule cards worked well in combination with the tools. As I walked around the table, I could see that confidence was growing.

When the dealer made the third pass, you could hear a pin drop.

"Dan?" Brenda asked. "What gives?" She held up one of her cards:

The story was the same for each player. There were two unintelligible cards in each hand. "Brenda, allow me to introduce Mr. Chan. He's from Malaysia."

"Maury, what were you saying about suckers?" Lonnie asked.

Maury threw his cards down on the table. "I'm not playing this game," he announced. The others followed his lead.

Mr. Chan handed the decks to the stranger on his right, who dealt out the cards faceup. Again, the familiar new tool and rule cards were topped by two unfamiliar cards.

"Allow me to introduce Señor Lopez from Brazil."

"I don't care who he is, I'm not touching those cards," Glenn said.

I nodded, and the third man dealt. The pattern repeated itself.

"I can't read cards like these. Where's this guy from?" Lonnie asked.

"Mr. Kemény is Hungarian," I said. "He loves to play the game."

"I don't care where he's from, I'm not playing with him," Lonnie declared.

Chan, Lopez, and Kemény shrugged, carefully divided the pot three ways, and moved from the dining room toward the front door.

"Wait just a minute," Maury almost shouted.

Mr. Chan turned and said, in perfect English with a slight New

England accent, "The game will be played with you or without you . . . good night."

And $7,000 walked out the door.

Interconnections

I gave the group a few minutes to calm down. "I might be able to arrange a rematch," I said.

"Great, but we've got to spend five years learning Portuguese, Malay, and Hungarian cards," Samantha groused.

"Those cards that you couldn't decipher were the same new tool and rule cards we've been playing with all along," I said.

"Could have fooled me," Doug replied.

"Language and culture don't affect the cards. The national barriers and boundaries have come down. My three friends were playing with the same cards that you are using, and that's one of the reasons they won. You didn't know that the decks were the same. In the United States our assumption was—and still is—that we are holding most of the good cards. This, of course, is yesterday's thinking."

I went to the flip chart. "Take a look a this."

Four Great Rivers of Change . . .

1) Technological innovation
2) Globalization
3) Decentralization of power and authority
4) Global demographic changes
(Note: 2, 3, 4 are tributaries of 1)

I said, "Imagine Earth's early history. Geophysical evidence indicates that many of Earth's present landmasses had huge bodies of water flowing over them like gigantic rivers. All rivers are in a constant state of change. The image is useful because accelerating change may, in fact, be the one thing that all the world's people have in common. It's as if humanity lived on the banks of the Mississippi, the Nile, the Rhine, and the Yangtze—the waters of which flowed

together, mingled, and continuously recirculated like the rivers of early history.

"Those who dwelt on the banks of the Mississippi might think that theirs was a unique existence. But in reality they were all drinking out of a universal—or at least, terrestrial—cup. Late-twentieth-century humankind isn't just drinking from a cup dipped into the four great rivers of change; we're up to our necks in the flood.

"We've talked at length about the first river, technological innovation, and with good reason. Technology alters reality. Without advanced technology, there would be no globalization, and we would not have the capability to manage the decentralization of power and authority. Global demographic changes would be greatly limited. Technological innovations therefore represent the enabling change agent, the greatest river of change, feeding into the other three. In 1989, Mr. Chan's country, Malaysia, became the third largest producer of semiconductors in the world. How was that possible? All you have to do is look at the rivers of change. Let's examine the other three.

"The second river, globalization, was barely a trickle until the dam began to break back in 1974 when Muhammad Ali fought Joe Frazier in Manila for the heavyweight championship of the world. It was a good boxing match, but as a sporting event it was minor stuff compared to the magnitude of the historic event that it symbolized. For the first time in history, a global event was simultaneously televised live around the world. Suddenly, this great river of globalization broke loose, releasing a flood of change that has swept the planet. Less than twenty years later the world watched combat of another sort, broadcast live simultaneously around the world—Scud missiles streaking across the night sky toward Tel Aviv and Riyadh during the Gulf War. We all watched life and death unfold as it happened, with both sides getting comprehensive information from live coverage of the event."

I pointed to the chart. "Decentralization of power and authority, the third great river of change, doesn't flow as majestically as the Rhine. It hammers along like Niagara Falls, taking out the Berlin Wall, the Warsaw Pact, the Soviet Union, and even washing the color out of Big Blue (IBM). The water level? Chest-deep and rising.

"And along comes global demographic change, the fourth great river: shifting youth in the twenty-first century to Asia and old age to the early industrialized nations of the world. Looking at the United States, this mighty river is redefining the terrain. For example:

1. Women born after 1960 will have more husbands than children.

2. The world's population heads for 6 billion by 1998.

3. Mexico City—not Tokyo, London, or New York—is the world's largest city.

4. Los Angeles is the second largest Chinese city outside of Asia, the second largest Japanese city outside of Japan, the largest Vietnamese city outside of Vietnam, and the largest Filipino city outside of the Philippines.

5. The United States of North America, the United States of Europe, and the United States of Asia begin to emerge.

"The four great rivers of change just keep on rolling."

Staying Close to Home

"I'll forgive you for sandbagging us on the cards just now, Dan, if you can tell me why I really have to worry about this trend toward globalization when I'm involved in the food products distribution business?" Glenn asked. "I don't have to worry about the Japanese, thank God. And this country is still the leading food producer. Are you saying that our status is being threatened?"

"Suppose that my answer is 'No, the United States will continue to be number one.' Does that make you feel better?"

"It makes me less apprehensive—yes."

"You're not worried that the British are the largest foreign owners of agricultural land in the United States and that they have begun buying into U.S. grocery store chains?"

"Well . . ."

"You don't care that your customers will have to turn to the competition to get specialty products sought after by their Hispanic and Asian clientele?"

"Well . . ."

"You don't care that Chile, Spain, Israel, and Indonesia are denting the Florida and California fresh produce market? Hawaii is already out of the pineapple-growing business. That whole trade has shifted into the western Pacific."

"I get your point," Glenn said.

"And you were the one with a vision of the Eiffel Tower," I pointed out.

"I was thinking of expanding the business over there, not having them come here."

"It's a two-way street."

"Make that a two-way river," Lonnie said.

"Two-way—but only recently," I said. "We've talked a good free-trade game in this country, but mainly because we didn't have all that much to fear from our competition. For the nineteenth century and most of the twentieth, imported products filled a vacuum; our huge internal market was the envy of the world, and demand far outstripped domestic capacity most of the time. There was enough business for everybody, particularly since the competition was on the other side of the Atlantic Ocean, which blunted the advantages possessed by Europe's established industries.

"When the market finally began to tighten, the answer was 'managed trade': Trade agreements were enacted to restrict imports of autos, machine tools, carbon steel, semiconductors, even motorcycles. It's another form of protectionism."

"What makes you think that pattern won't continue to hold?" Doug asked.

"The pattern would continue if it weren't for the four rivers of change. Our leverage is eroding. A little over ten years ago, the president of the United States was president of the Free World. Today, he is president of the world's largest debtor nation. We depend on the continued goodwill of Japan and Europe to invest in the United States to finance our deficit. When there was a war to fight in the Persian Gulf in 1991, the U.S. secretary of state first had to go on a fund-raising mission.

"I believe that the long-running stalemate in the Uruguay round of GATT trade talks on agricultural issues would have been resolved in our favor during the Bush administration if it had occurred ten or fifteen years ago, simply because the Europeans couldn't have risked getting us riled up."

"But, Dan, those farm subsidies are outrageous," Tanya said.

"Of course they are. However, if we were taking the long view, I'd want those subsidies to stay in effect. Why? Because they'll bankrupt my competition by the end of the century. Meanwhile, we should be cultivating open markets in Latin America—Brazil is moving in that direction. Why fight to sell soybean oil to the French and Spanish? They're awash in olive oil. Feed sub-Saharan Africa and get an entire continent back on its feet as a viable market."

"I think we should stay closer to home and feed and shelter our own people," Brenda said. "As an African-American, I think we should do more for starving people in Africa, obviously. But there are limits."

I pointed to the chart. "Globalization, Brenda. We are staying close to home. The day the bell sounded for round one in the Frazier–Ali fight was the day that home became planet Earth. Culture, commerce, and political interaction all depend on communication. Not intermittent communication—instant and constant communication. The historical expansion of the social unit from family to tribe, from tribe to village, town, city, state, region, nation—continent—was a function of technology, principally communication technology, followed by what is called the life-support technologies of food, medicine, and finance.

"The Soviet Union fell apart because there was little or no up-to-date technological glue holding it together. The empire lasted as long as it did because of police state tactics, inertia, and military force. The pieces are now being drawn into new patterns by the powerful effects of twenty-first-century technology."

"I disagree with you about Africa, Dan, but Eastern Europe and the former Soviet Union have a lot more promise as markets," Maury said. "Still, it may be decades before things stabilize. Businesses are going to have to be extremely cautious and take a wait-and-see attitude."

"Maury, thank you," I said. "You've set the stage for me to introduce another key concept." I went to the flip chart.

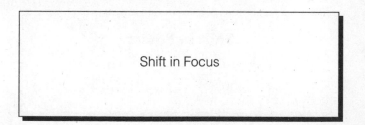

Shift in Focus

"When I was child, I found an old magnifying glass and started to play with it. Before long, I discovered that I could concentrate the light to shine in a beam that was very bright and hot. To my amazement, I could actually burn holes in dried leaves. It taught me the power of focus. If you can concentrate your attention and energy on one spot, great power is released.

"You can also burn an ugly hole in your parent's carefully manicured lawn, as well," I added. "After reseeding the lawn, as I mulled over the discovery I realized that the ability to focus on a problem or a goal is crucial to success. If you've ever watched a small child examining a seashell, you've seen the power of focus in action. He or she is totally absorbed in learning everything there is to learn about the shell—where it came from, what it's made of, how it tastes. The child is focused on figuring out what makes the world tick. Great geniuses are childlike in that same way. A story is told about Richard Feynman, the pioneering physicist, who was so interested in the simplest things around him that he would stop and closely examine the worn granite steps leading to the front door of the Princeton University Library. He could actually picture the individual molecules of granite being scraped off by the feet of generations of students. The power of focus.

"But just like the beam of light through the magnifying glass—moving from the dried leaf to the blade of grass—there is now a great shift in focus taking place. As individuals, we do it by directing our attention elsewhere: Feynman decides to look at the shingles on the roof of the library building. Likewise, evolving technology also changes the focus. It acts as a moving magnifying glass.

"The reason I got started on this tack was Maury's statement about taking a wait-and-see attitude toward Eastern Europe and the former Soviet Union. Because of technological changes, there has been a shift in focus.* The magnifying glass has moved on. I'll show you on the flip chart."

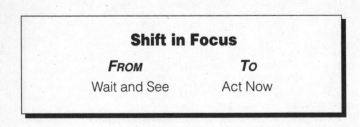

Shift in Focus

FROM	TO
Wait and See	Act Now

*See Appendix D.

I continued, "Changes are happening too rapidly for wait-and-see. Imagine a time line to cover the last 2,000 years, with one inch representing 100 years and Xs marking the introduction of breakthrough technologies. Roughly 98 percent of the breakthroughs would be crowded into the last inch. What does that tell us? The impact of technology will intensify, and we must act.

"The previous nineteen inches, with the orderly march of Xs spread across the decades, are just right for wait-and-see. The Germans, for one example, aren't looking at that part of the time line, and they aren't waiting. They took a precipitous plunge into reunification, knowing that it wasn't going to be easy. However, in ten years, or less, the worst is going to be over for the Germans; they will be stronger than ever economically and socially. Acting now *will* pay off later.

"On Yugoslavia, however, the United States and its NATO allies chose to wait and see. The result—horrible civil war that spread instability throughout the Balkan region. Wait-and-see doesn't seem to work!

"I wonder if the Germans have read the American historian Henry Adams and his theory of acceleration. It really isn't a theory as much as it is an astute observation. Back around 1900, Adams observed that things were happening much faster than they had in the past. The acceleration, or Adams's noticing it, coincided with the Gilded Age, the zenith of the industrial epoch in the United States. In the years since, we've gotten so accustomed to the acceleration that we hardly even notice it.

"Let me show you a fun exercise to squeeze 50,000 years of human history into 50." I went to the chart again.

Time Compression Chart

50 years ago	Cave man
5 years ago	Communications with pictures
	(Not Hollywood—the walls of a cave)
6 months ago	The invention of the printing press
1 month ago	The electric light
3 weeks ago	The Wright brothers' first flight
1½ weeks ago	The first TV set sold
Yesterday	Berlin Wall fell

"Like the passengers on an accelerating train, we focus on a passing barn or meadow for a moment and then must refocus on a stand of pine trees or a cluster of houses. As the train goes faster and faster, we can lean out the window and try to hold on to the vision of the barn, but in seconds, it's gone. The operative questions are: 'What am I focused on? Am I still trying to focus on the barn?'"

I wrote on the flip chart:

Current Management Focus

Manage by control
Decision by command
Negative reinforcement of negative behaviors

"Now the shifts are added to the right of the chart."

Management's Shift in Focus

FROM	TO
Management	Leadership
Control	Commitment
Decision by command	Decision by consensus
Negative reinforcement	Positive reinforcement

"I follow you, Dan, as far as the effects of the acceleration of time—or the compression of time, for that matter—but why this particular set of shifts?" Doug asked.

"Throughout these games I have been helping you to go beyond your competition. This requires innovation—not imitation. It is important to know that there are two types of innovation that can help us. One of them is easy to identify: technological innovation. The other isn't so easy. I use the rule cards to get at the essence of what could also be described as organizational innovation.

"Both technological and organizational innovations must be orchestrated together. If you concentrate on one without the other, you won't stay beyond your competition for long. Let's explore organizational innovations by looking at the shifts in focus they can cause."

Corporate Culture Shift

FROM	TO
Cost/growth/control	Quality/innovation/service
Corporate groups	Partnerships
Periodic improvement	Continuous improvement
React to change	Initiate change

"What we're looking at on the right-hand side of the chart is a reflection of organizational innovations in response to technological innovation. Look at the effect on the area of human resources."

Human Resources Shift

FROM	TO
Focus on task	Focus on process
Individual work	Teamwork
Individual values	Shared values
Upgrading technology	Upgrading people
Job security	Job adaptability
Promote by seniority	Promote by performance
Retirement	Re-engagement

"If these shifts in focus do not take place," I said, "the technological changes are sabotaged. It is one of the primary reasons for the failure of attempts to introduce cutting-edge technology. And the opposite is also true: Organizational innovation without the aid of technological innovation is often destined to falter or fail. That's what happens to many companies that are structurally driven to the point that the composition of the organization is its central focus. IBM is an example of a company that is organizationally focused in this way. The organizational structure is everything. Corporate culture is defined by it. When trouble sets in, the response of an organization-based company is to reorganize. IBM, unfortunately, has been fighting three major areas of difficulty at the same time: an obsolete organizational structure, an obsolete corporate culture, and a maturing market for their biggest cash cow. Focusing on reorganizing again and again will not stop the downward slide. An innovation-based company, on the other hand, uses technology to change the rules of the game."

"So, where and how do we shift the focus technologically and organizationally?" Doug asked.

Global Lip Service

Before I could answer the question, Tanya looked up from a pile of papers spread on the table in front of her. "I found it," she said in triumph. "You used a term a while back that I wrote down. It seems apt for the shift in focus. May I write it on the chart?"

"Be my guest."

Tanya moved to the tripod that held the chart and carefully removed the top sheet. She took the marker and wrote:

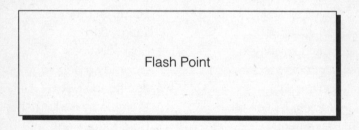

Flash Point

"The Communication Age reached an application flash point, the Frazier–Ali fight," Tanya said. "That flash point brought on globalization."

"Tanya, if I were Groucho Marx, a gong would sound and a duck would drop down from the ceiling. You've said the magic words."

"I get it," Lonnie said. "The core technology, the availability of applications, and then when it hits the right price, all jelled and boom!—flash point. Now, there's got to be a shift in focus whether we like it or not."

"Except I don't think things that jell ever boom," Samantha said.

"That's what's known as a shift in metaphors," Brenda added. "A high-tech version of a mixed metaphor."

Lonnie shook his head. "I make a dynamite cranberry dessert. I call it Cranberry Bomb."

"Can we suspend the wordplay?" Maury asked. "I'd like to get back to the Age of Globalization. Something tells me it is the only way I'm going to retrieve my blank check from three guys named Chan, Lopez, and Kemény."

I went to the flip chart and underlined Tanya's words to remind the group of the importance of flash point and then turned the page.

Shift in Focus Globalization

FROM	TO
Thinking global	Being global
Domestic market	Global market
Global competition	Global collaboration

"Before the flash point, American business thought a lot about the global market. Now that the flash point has occurred, it is time to shift the focus and actually *be* global. Not just the big guys; they're moving fast in that direction because they have the capital resources. Maury's gone international."

He nodded. "Yes, but I have the darnedest time convincing my best executives to take overseas assignments," he said. "They're afraid that they're leaving the fast track. I've got an executive in Austria who I think I'll bring home and promote over the heads of six other more senior guys. His international experience is worth a lot to the company. That might be an effective wake-up call."

"You've got a typical problem, Maury," I said, "and the solution is right on target. The global track is the fast track for individual careers and businesses, small and large. I want small and medium-size businesses to also put on their running shoes and enter this Age of Globalization.

"Samantha? Lonnie? If you're not global, you're not playing with all the new tools and rules."

Samantha: "It never occurred to me that I needed to shift my focus overseas. But what if the French or the Italians decided to steal our cash cow—life insurance? The federal government probably wouldn't lift a finger to stop them."

Lonnie: "I'm going to start contacting international companies that have videoconferencing setups, and I'll bid on their projects. I could sit in my office and actually show the prospective clients the plans and go back and forth making changes. Maybe I can find a way to coat the Taj Mahal with diamond thin-films."

That drew a healthy laugh from the group, the first since their run-in with the three strangers. But Glenn pounced on the Taj Mahal idea.

"That's not so crazy, Lonnie," he said. "I don't know about the Taj Mahal, but all those European monuments and artifacts are literally melting because of the air pollution. The Italians would love you. All roads lead to Rome, after all."

"Leonardo da Vinci, here I come," Lonnie said.

Border Crossings

"Back to the shifts in focus," I said. "We need to shift from paying lip service about being global to actually doing it. Similarly, the focus on the domestic market needs to shift to the global market. Why? Because technology has abolished national borders. Ali–Frazier was only the first of many flash points. The cumulative effect has been the meltdown of the nation-state. What country the corporate headquarters are in is becoming meaningless. Hypermobility has been injected into a business milieu that was once static in the extreme."

World War III

"The final shift, from global competition to global collaboration, hit one of its flash points with the formation of business partnerships between U.S. manufacturers and Japanese manufacturers in the 1980s," I said. "American businesses like GM, Ford, and Texas Instruments began lining up joint arrangements with Japanese firms—Toyota, Mazda, and, in the case of Texas Instruments, Sharp. Those companies saw the advantages of collaboration. No one corporation is strong enough to go it alone in this Age of Globalization. There must be creative combinations of strengths and specialties and market expertise. Even the large U.S. consortium of semiconductor manufacturers, Sematech, is going to have to reach out for input from foreign companies to be truly effective.

"It's time we realize that World War III has already started, but it's a different kind of war. The purpose of a war is to win," I said. Everyone nodded in agreement. "It's hard to determine who wins when you use the old tools and fight by the old rules of warfare. The new weapons of warfare are to attack with quality products and services that the other nations' people can't resist. All warring nations win. Each warring nation will help the others to fight because it raises the standard of living for all sides. Wealth is not distributed, it's created. By increasing the pie, there's enough to go around. It is a confusing war. Our friends are our enemies and our enemies are our friends. The alliances and adversarial situations are going to be constantly shifting. All sides do win, however, and yet there are losers."

"Who loses?" Lonnie asked.

"The industries, companies, and individuals who are playing with the old deck and playing by the old rules," I said. "In this case, war is not just a metaphor; it is an accurate description of what's going on. Fifty years ago, the Japanese invaded foreign territories, set up bases of operation, and shipped money, natural resources, and goods back to the homeland. They are doing the same thing today without firing a shot. I'm not bashing them. They've recognized the technological and organizational flash points and shifted their focus away from the military, and they are now reaping the benefits of peaceful economic warfare. The Germans are doing the same thing, as are the little Dragons of Asia: Singapore, Hong Kong, South Korea, Taiwan, and Malaysia.

"If we don't wake up, mobilize, and start fighting our own peaceful economic war, we are going to be vanquished. I'm not saying that the United States will be under armed occupation, but our financial and social strength will be drastically weakened. Over the last twenty years there has been a decline in U.S. living standards—and we haven't seen anything yet if we don't implement that first shift in focus that I gave you."

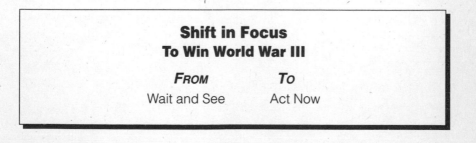

Shift in Focus
To Win World War III

FROM	TO
Wait and See	Act Now

"When the Japanese attacked Pearl Harbor, the United States did not wait and see what would happen next. Act now was the only answer. Roughly forty years later, the disaster of Pearl Harbor proportions in the peaceful economic war had nothing whatsoever to do with the Japanese. We bombed and strafed our own economy by allowing the deficit to balloon and push the national debt to $3.8 trillion (as of 1992). The numbers don't even convey the enormity of the disaster. To understand how much a trillion is, think of it this way. If you had a stack of one-thousand-dollar bills totaling $1 million in front of you on this table, it would be about three inches high. A billion dollars in one-thousand-dollar bills would reach about 250 feet. One trillion dollars would be a little over forty-seven miles high. To pay off the debt in 1992, every American taxpayer would have had to cough up an estimated $47,000."

Replay

"Sorry, I'm a little short. I'll have to ask Chan, Lopez, and Kemény to stake me," Lonnie said.

"Turn around," I said. The group just sat there staring at me. "Turn around," I said again. Finally, all heads swiveled toward the doorway.

Chan, Lopez, and Kemény were standing there, smiling.

"Ladies and gentlemen," I said. "It's time to play poker."

Samantha had the last words: "And take no prisoners!"

Conclusion

Going Beyond Your Competition

Glenn poured the champagne, and Brenda proposed a toast. We lifted our glasses. "To using new tools to change the rules of the game," she said.

Everyone took a healthy slug of the Moet-Chandon.

It had been a civilized finale. I had done the dealing and used the decks of rule and tool cards that had become second nature to all the players. Chan, Lopez, and Kemény turned out to be friendly and skilled players whom I had taught the new card game while attending an international seminar the previous year. They weren't looking for competition. Collaboration was the name of their game.

I picked up my glass. "My toast is this: to going beyond your competition." As we touched our glasses, it was obvious that the feeling among the players had changed during the games. There was a sense of confidence, purpose, and focus that would reach far beyond our games. "Understanding the significance of major technological trends and using this knowledge to go beyond your competition is what this has been all about," I said. "I have used a lot of different metaphors to give you a vision of the future: sports, trains, planes, beams of light, cards, and harshest of all, war. Doug put his finger on it when he told me that sports and war are the central metaphors of American life. It's little wonder, since we've been at war for most of the twentieth century, and for most of the same period sports has been our other national pastime.

"Sports are here to stay, but fighting wars in traditional ways is rapidly becoming obsolete. It will be important for the advanced nations of the world, including the newly advanced nations, to help the others learn to attack each other with quality products and services instead of bombs, guns, tanks, and missiles.

"This intense study of technology that I have made over the last decade has turned me into a genuine optimist, but not a blind optimist. I can surely provide a long list of major problems. My optimism is based on the potential good that can be done if we focus new technology in the right direction. We have the ability to alter reality on a global level. We all must jump into the Communication Age and help world leaders see the new big picture to enable them to shift from crisis management to opportunity management. Gorbachev did it, the Chinese will do it, the United States must also do it. The key is to think beyond personal and national ego, which are destructive attitudes, and then define ideal goals. I feel good about the fact that the old technology of war does not produce clear winners anymore and has just about run its course.

"During the Cold War, the USSR, China, the United States, and a few others developed their war-making capacity to such a high level of efficiency and sophistication that the technology is now virtually unusable except perhaps by a madman. The potential destruction of our nuclear arsenal is too horrible for sane people to countenance. Clearly, the devastation would be so complete that it would be hard to determine a winner. If you were the leader of China, would you rather have a billion hydrogen bombs or a billion highly literate and retrainable people? I know what I would want because only one of the two options will produce a definable winner.

"There will be the occasional two-bit terrorist, and ethnic strife will flare up starting smaller but deadly wars. After all, sometimes it's easier to hate than to change. We are all flawed humans who make mistakes. My hope is that by 2025, all nations will have learned to go beyond deadly war to fighting win-win economic wars. Hand-to-hand combat is best left to sports like football and rugby.

"Why do I think this is actually possible? There's nothing to be gained from modern war. Technology guarantees that the losers pay a fiendishly high human cost. The winners practically go bankrupt. It's not even good politics. In the end, President Bush achieved no lasting political benefits from the Persian Gulf War and may have been hurt politically because he pulled back from inflicting a level of destruction on the enemy sufficient to topple Saddam Hussein. If he had not shown restraint, the carnage would probably have sickened even many of the most hawkish advocates of that conflict.

"Glenn, in the eighties, Iraq and Iran fought for eight years. Tell me who won?" I asked.

He hesitated before answering. "Iran?"

"No, Iraq. It's hard to remember, isn't it? If you look at the two countries today, it's still hard to tell.

"Brenda, who won the war between Vietnam and Cambodia?"

"Pol Pot, the Cambodian communist leader."

"No, Vietnam."

"I thought the movie *The Killing Fields* . . ."

"That was before Vietnam invaded Cambodia and chased Pol Pot into hiding. And since I've mentioned Indochina, North Vietnam won the war in South Vietnam, and the victors have been living in abject poverty ever since.

"As you can see, keeping score is next to impossible—and winning often means losing in the long run. What's the point of having a war then? I'd dare say that in ten years few will remember who won the Yugoslavian civil war. Talk about no-win situations. Integrating socially and economically with the rest of Europe would have been a better tactic for all sides."

Taking the Profit out of War

"The rule *If it works, it's obsolete,* certainly applies to war. The technology of mass destruction works too well. An estimated eighty million people in Third World countries have died as a result of military conflicts in the forty-year period of the Cold War. That's the bad news. The good news is that without the old ideological engine to drive the killing machine, the momentum will subside. The lucrative international arms trade can only live a limited time off the fat accumulated during the Cold War. That peculiar and grotesque form of free enterprise is a spin-off of old-fashioned superpower rivalries. Can it exist without superpowers? Probably, but only in a very reduced form. Soon the industrialized world will run out of excuses for not converting many of its militarized manufacturing facilities to civilian uses. It just won't be good business anymore. Making tanks and bombers is expensive, and without the deep pockets of the former Soviet Union and the United States, the profits will be meager. If they ask themselves, are these the tools for fighting? the answer might be yes; but if they ask, are these the tools for winning? the answer would be no."

"A world without shooting wars?" Maury asked.

"There will be occasional hot spots. Iran bought old submarines from the Russians in 1992 and will probably try them out sometime. Getting the Kremlin's military hardware off the 'street' will be a major diplomatic challenge. But the biggest potential troublemakers have woken up to the fact that they can fight and win wars without submarines. There's too much to lose in a real shooting war, which is precisely how technology has made war increasingly obsolete. Those who have nothing to lose will probably get into confrontations from time to time. But it's in everyone's best interest to take steps to see that no nation or region is so marginalized that war is the only means of competing for wealth and power. I think it was Abraham Lincoln who said, 'The best way to defeat an enemy is to make him a friend.'"

As I talked, I had been circling the dining room table. I moved back toward my flip chart. "I can make a prediction that I'm confident will come true. For each of you here tonight, the next fifteen years will be the most exciting years of your lives. We've never had opportunities like we do today. But to seize them we have to be opportunity managers setting ideal goals.

"Crisis managers are still in charge. That has got to change. How else can we explain that, as of 1992, there were 700 research laboratories in the United States spending $23 billion a year on military-related R&D. General Rip Van Winkle went to sleep and missed Gorby, Glasnost, and the Golden Arches in downtown Moscow. We need our military and our R&D labs to help us fight the new economic war."

"Not all of that is wasted," Lonnie said. "Eventually that R&D will spill over into the civilian sector."

"Unfortunately, this has not been the case very often," I responded. "Research is a funny thing. Government statistics reveal that less than 10 percent of the money spent on defense R&D has ever resulted in commercial products. Much of the work that does not have direct military application sits on the shelf until somebody almost literally bumps into it. Just to see how long it takes, I'm watching one breakthrough discovered by the air force a few years ago: using microwaves to 'boil' toxic spills out of the soil. The commercial possibilities are very promising, particularly when you think of all the old gas stations that had rusting underground tanks that leeched fuel into the soil. Removing the old tanks does not remove the problem. Many railroad switching yards have had the same situation with chemical spills. But the air force isn't in business to clean up gas stations and rail lines, nor is it interested in spreading the word about its findings, many of which are immediately stamped Top Secret. It's one of the reasons I publish a technology newsletter. The people who are paying for the R&D should at least know how their money is spent and what's available.

"Doug, you'll appreciate this. Here's the traditional flow of technology produced by our investment in R&D." I went to the flip chart.

The Trickle-Down Flow of New Technology

First it goes to the Military
Then to Medical Facilities
Then to Business
Then it's made into Toys
Then to Education
(*But with no instructions*)

"Does this fit the past or the future, Doug?"

"The distant past. Sometimes I think the trickle never gets past the toys," he said.

Five Words

"What you're saying about going beyond competition sounds good," Maury said, "but I have my doubts. I really think we're heading for a period of even more intense competition as this new technology trickles down, to use your words."

"You'd really like to have a monopoly on the latest technology, wouldn't you, Maury?" I asked.

"Yes, I have to admit it. I would."

"Those days are over. There will be no monopolies on technology. But I'll give you a monopoly that's even better. I don't know what to call it. Since making up words is in vogue these days, how about *flexigration?* It's a rough amalgam of four key words that we've already discussed: communication, flexibility, integration, and orchestration. And I've added a fifth word, anticipation.

"Here's how to turn the five key words into action."

---

1. Anticipation

   a. Using the core technologies to anticipate permanent changes

   b. Using the 24 cards to anticipate changes and opportunities

   c. Using demographic changes to anticipate opportunities

   d. Using dynamic (electronic searching of data bases) information to keep up on change

---

2. Communication

   a. With all employees

   b. With our suppliers

   c. With our distributors

   d. With our manufacturers

   e. With our sales force

   f. With our customers

   g. With the public

3. Flexibility

   a. In what we make
   b. In how we advertise
   c. In how we think
   d. In how we view our market
   e. In how we view our competition
   f. In how we view our product
   g. In how we view the business we are in
   h. In the policies we have
   i. In the agreements we have and make

4. Integration

   a. Of departments to work in harmony
   b. Of information to allow access to all of our employees
   c. Of strategies to give focus
   d. Of people and their values
   e. Of goals to show the relationship

5. Orchestration

   a. Of the application and use of technology
   b. Of all the various parts and players to form a symphonic harmony
     (Business as an art form)

"I like all five words, but my favorite is *orchestration*. It puts the individual—you and me—in charge. We're up there on the podium, baton in hand, shaping the music—shaping the future.

"Brenda, I'm going to ask you a question, but I don't want you to think about the answer. I want a quick, off-the-top-of-your-head response."

"Fire away."

"At work, are you a leader?"

"No. I do my job, and I'm pretty good at it."

I nodded. "When businesspeople are surveyed, and when that same question is asked, the overwhelming majority answer just the way Brenda did. Very few people regard themselves as leaders, and even many of the most successful top executives feel that way. I don't know what it is, but there is almost a stigma attached to leadership if you're not in a senior executive position. It may be a reflection of our antiauthoritarian tradition. All men are created equal, and therefore no one is in charge. There are no leaders. Consequently, as managers advance to executive positions, they keep managing. They're running our business enterprises without an essential ingredient—leadership. True leaders are opportunity managers, while managers tend to be crisis managers who are often cold, impersonal, and by the book. They say, 'I manage people and processes.' They should be proclaiming, 'I orchestrate opportunities.'

"The next time you vote for president, ask yourself which candidate is capable of orchestrating change. Evaluate your boss and the company you work for by asking how effective the orchestration is. Size up your own performance by asking if you're orchestrating your own future the way you should."

Winners and Losers

"I can tell that Maury is skeptical about the concept of going beyond competition. He thinks that competition is going to intensify. He's right. It will intensify—for the competers. For instance, if the United States competes for low-wage jobs with Mexico and the rest of the Third World, we lose by winning a Third World standard of living. If we compete for the export market without collaboration with domestic and international partners, we lose. If we compete in the manufacturing sector by cutting prices, we lose. If we focus on imitating the leading country or manufacturer, we lose."

I turned the page of the flip chart.

Focus on Competition Has Always Been
a Formula for Mediocrity

"Our word *mediocre* comes from combining two old Latin words into an idiomatic expression that is literally translated to mean 'halfway up a mountain.' The idea behind the expression is that a person is stranded in a no-man's-land, halfway between two ideals.

"That's exactly what a preoccupation with competition does to any business or person. Consider a few of the more obvious conditions that emerge when a company focuses attention primarily on its competition."

The Hidden Cost of Competition

1) It causes a loss of identity. All the competitors in a given arena tend to copy each other's products, services, and methods to the point where it's hard for customers and employees to distinguish one from the others.
2) It causes individuals and firms to get so caught up in struggling with each other that they lose sight of their own customers and markets.
3) It forces so much concern over minor cost-reduction measures that innovations that could create major savings get completely overlooked.
4) It creates so much concern over short-term sales and profits that long-term profits, problems, and opportunities consistently get overlooked.
5) It so limits the resources available for innovation that change can come about only as a result of a crisis.
6) It lulls companies into a complacency born of the belief that they are secure as long as they are currently maintaining their share of an identified market.

"Perhaps there was a time when it made sense to play the one-upmanship game of keeping up with the competition. But the dramatic changes spawned by science and technology have made that a perilous game for the present and a formula for disaster for the future.

"Just the way technology has changed reality, making traditional war obsolete, technology and the rapid pace of change that it has generated have made competition obsolete. It's too dangerous to compete. We have too much to lose to compete. I want a sure thing. And the only way to get it is to adopt a personal and business strategy for integrating technology and people for the purpose of achieving our ideal goals.

"I'll write that sentence on the flip chart," I said.

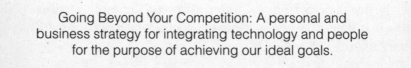

Going Beyond Your Competition: A personal and business strategy for integrating technology and people for the purpose of achieving our ideal goals.

"You've come up with a formula that you call going beyond your competition, which amounts to a way for making our ideal dreams come true. Our ideal dreams?" Tanya asked with a smile.

"Have you heard Natalie Cole, daughter of the late Nat King Cole, sing with her deceased father?" I asked. Samantha and most of the others nodded yes. "Beautiful, wasn't it?" Again, a yes was indicated. "That was impossible, until a new tool of technology allowed her to separate her father's voice from the rest of the sounds in the old recordings. A new tool allowed her to achieve what was to her an impossible, *ideal goal*. She used a new *tool* to change the *rules*. She created a new *role* for herself as an entertainer and achieved a new goal. New tools, therefore, allow you to achieve new goals, redefine your role, and change the rules of the game. For her, it meant millions of dollars, and she forever changed entertainment. She went beyond the competition.

"Tanya, your dreams are truly beyond competition if you keep dreaming and keep realizing those dreams. Nobody can come close to competing. It's when we try to live someone else's dreams, or live dreams from the past, that we make ourselves vulnerable to bankruptcy, unemployment, and unhappiness.

"One of my favorite statements about the future is from Antoine de Saint-Exupéry, the French writer and early aviation pioneer. He wrote *The Little Prince* and other wonderful books about flying. Those of us who love to fly, and I spent years working with experimental aircraft, are actually living one of humankind's oldest dreams—the dream of flight. In the cockpit of a plane at 30,000 feet, you can understand how dreams come true. It makes you a futurist, through and through.

"Saint-Exupéry said, 'As for the future, your task is not to foresee, but to enable it.'"

Taking Stock

"To become enablers, you'll need to use these blank triangular and octangular cards I'm about to give you. Write your challenges on the triangle cards and the old rules of success you follow on the octagon-shaped cards. When you play as we did, you will find solid strategies for shaping your future.

"Now, I'll give you five questions that will help you determine if you are going beyond competition."

Are You Going Beyond Competition?

1) To what extent have your current products, services, or career approaches been shaped by your competitors?
2) In what significant ways has your field of expertise changed during the last decade? Has that affected the way you do things?
3) What would you (as an individual or company) do differently if you had no competition? Could doing some of those things make you the competition that everyone else tries to keep up with?
4) Do you keep plodding along day after day, doing essentially the same things you've always done, or do you pause regularly to get a sense of the new big picture?
5) What percentage of your time, energies, and resources do you now invest in education and upgrading your skills? Is that an adequate amount?

Lonnie held up his hand to be recognized for a question. "On that last item, number five. Can you give us a ballpark percentage on how much we should be investing in upgrading our skills?"

"One number wouldn't apply to everyone," I replied. "Approach it this way. Look back over the last year and add up the number of hours you spent on personal re-education. If you're honest about it, the figure will probably be shockingly low. Resolve to double that figure—square it—in each of the next five years."

"Wow! That could turn out to be a full-time job," Glenn observed.

"But what a great full-time job" was Brenda's immediate comeback. "What do you do for living?" she asked rhetorically and then answered: "I invest in myself."

Empowerment and Other Buzzwords

I held up a stack of envelopes. "As a last exercise, I'm going to hand out ten envelopes to each of you." I passed them around the table.

"There's a story to go with these envelopes. A friend of mine was hired as the CEO of a large but shaky corporation. His predecessor left him a bottle of Scotch and ten envelopes. There was a note with them that said, 'When you get into trouble open the first envelope.'

"Things went well for a few months, but inevitably a disaster struck, and my friend opened the first envelope. There was a card inside; on it was typed the admonition—'Re-engineer everything!'

"That sounded like a good idea, so my friend re-engineered everything. For a time his problems seemed solved. Again disaster struck, and advances in technology rendered his new processes and facilities obsolete. He opened the second envelope. The card said, 'Establish a learning organization!' My friend started a company library filled with business books and opened an in-house school for all employees.

"Time passed, things improved, but the company's competition must have done the same thing because profits started to flatten again. Then there was a shift in the marketplace, causing another crisis. He opened the third envelope. 'Empower your employees,' the card directed. The next day he wrote a memo empowering his employees to take action and solve problems.

"The employees loved their new power and got to work, but there didn't seem to be a focus to their actions. Momentum increased, but the direction was unclear. My friend opened envelope number four, figuring that this time his predecessor's advice would surely pay off. 'Benchmark!' He reviewed the best of his competition's products and copied the best ideas. At great expense he gained a little market share.

"Envelope five was opened. 'Go with business process management.' He gave it a try, but he needed more, so he opened envelope six. 'Form autonomous, self-directed work teams.' This sounded good so he did just that. Unfortunately, communications broke down, and money was being spent solving problems that other teams had already solved.

"Envelope seven. 'Total Quality Management.' My friend decided TQM had to be the answer. He told the employee suggestion systems management people to forget the suggestion boxes and report to the TQM team leaders. Everyone focused on quality. This did improve overall quality by a great deal, but the competition, the Japanese, had already done this ten years ago and were focusing on flexible manufacturing. They were still eating his lunch.

"Envelope eight. 'Strategic Quality Management.' That sounded even better. It didn't help.

"Envelope nine. 'Quality Circles.' He liked TQM so he tried it. It also helped, but the competition continued to run circles around him.

"The last envelope! My friend was absolutely desperate. He opened it and read the card: 'Prepare ten more envelopes.'"

"And drink the Scotch," Lonnie said.

"That was gone long before the final envelope was opened," I said. "All of those fashionable business 'movements' or concepts are great. They'll show results. But remember, your competition is reading the same books and doing the same things. You will be working very hard just to keep up. Without taking the new big picture into account, without a strategy integrating technology and people for the purpose of achieving *ideal* goals, and without a focus on innovation, it's impossible to go beyond the competition and stay there.

"I'm worried about all of you. While you were learning to use the new cards, your desk filled up with crisis upon crisis. You'll go back to the office tomorrow and get swept away, looking forward to getting back to normal—status quo.

"Remember, status quo is now defined as rapid change.

"You need to make a commitment to your future. In this high-tech world where we are all plugged into our pagers and portable phones, there is a time to plug in and a time to unplug. You need to unplug at least once a week for at least an hour. During that time, you need to look at the predictable problems you will be facing in the future, based on the new cards, and solve those problems before you have them. This is the only way to keep your desk clear in times of rapid change.

"So when you get back to the office, close the door and open the first envelope."

I pointed at the envelopes in front of the seven players. "Go ahead. Open one now. Why wait?"

"It's empty," Samantha said.

"They're all empty. Fill them with your ideal goals, fill them with your futureview, and use the new tools and rules to go beyond your competition."

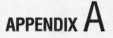

The Taxonomy of High Technology

Twenty Core Technologies Shaping the Future

1. Genetic Engineering

All living organisms are made of cells, and in those cells are genes that have a readable code defining all aspects of the plant or animal. Genetic engineering techniques allow scientists to eliminate or enhance a specific trait.

2. Advanced Biochemistry

Using advanced biological techniques, biochemists are creating new disease diagnostic systems, highly effective "superdrugs," advanced drug delivery systems, and a variety of new bioindustrial applications.

3. Digital Electronics

Digital devices translate signals into the 0s and 1s that computers understand. The key advantage of digital over analog is the ability to generate, process, store, and transmit *all* forms of information, such as data, text, sound, and images from one device to another. Traditional electronic analog devices, as well as magnetic and optical devices, can use digital techniques.

4. Optical Data Storage

Optical memory storage systems use lasers to read information that is stored in digital form. They can contain data, text, sound, and images in any combination and can randomly access the information at relatively high speeds. Another key benefit is the ability to store large amounts of information in a small amount of space.

5. Advanced Video Displays

There are two main types of advanced video displays: advanced flat-panel displays, which will provide us with full-color, flat, and lightweight television screens in a variety of sizes; and high-definition television (HDTV), which describes a very high-resolution screen whose dimensions resemble those of a picture in a movie theater.

6. Advanced Computers

Computers are electronic calculating machines that can process information and follow programmed instructions. Advanced computers cover all related hardware and systems that are based on advanced chip technology such as hand-held and personal computers, workstations, and supercomputers.

7. Distributed Computing

Distributed computing includes both enterprise and multiple-enterprise computer integration and the transparent multiuser sharing of information and applications across a multivendor computer network that supports day-to-day business activities.

8. Artificial Intelligence

Artificial intelligence is the capability of a computer to perform functions that are normally attributed to human intelligence, such as learning, adapting, recognizing, classifying, reasoning, self-correction, and improvement.

9. Lasers

The word "laser" stands for "light amplification by stimulated emission of radiation." Laser light covers a narrow range of wavelengths, tends to be coherent, and is emitted in a narrow directional beam of high intensity. Laser devices can range in size from a pinhead to a football field. Their light ranges from invisible ultraviolet and infrared through all colors of the rainbow.

10. Fiber Optics

Fiber optics provide a digital highway in which photons, or particles of light, travel. An optical fiber is a hair-thin strand of glass composed of silicon and other materials with a light-transmitting core and a layer of material that keeps the light from straying. They are the fastest, most reliable way to send information. To upgrade a fiber-optic-system only the senders and receivers have to be changed, not the fiber. They can carry more than 100 times more information and are easier to upgrade than conventional copper wire; they are impervious to temperature extremes, corrosion, and electrical noise.

11. Microwaves

Microwaves are electromagnetic waves having a wavelength in the region between infrared and shortwave radio. Currently, microwaves have two major application categories: sending wireless digital information and heating objects by causing molecular movement inside the object.

12. Advanced Satellites

As advanced satellites with diverse uses are put into orbit by more and more countries, they will play an ever increasing role in worldwide government and business communications as well as in studying, mapping, and surveying the earth.

13. Photovoltaic Cells (PV)

When photons of sunlight strike a solar cell, electrons are knocked free from silicon atoms and are drawn off by a grid of metal conductors. This action yields a flow of electricity (direct current). PV cells require no fuel and are nonpolluting. They are self-contained with no moving parts and have a lifetime of more than twenty years.

14. Micromechanics

Micromechanics involves the designing and building of tiny mechanisms, such as valves, accelerometers, pressure and force sensors, and surgical tools. Micromachines can be etched in batches on silicon wafers and then sliced into separate chips. In addition to silicon, they can be made out of metals such as nickel.

15. New Polymers

Polymers are complex chemical structures that can be combined with reinforcing substances and adapted to many uses. By rearranging loops and chains of carbon, oxygen, hydrogen, and nitrogen, chemists are producing polymers that can conduct electricity, dissolve in sunlight, carry light waves, and function as moving parts in automobiles. Currently, there are over 60,000 different polymers with applications ranging from garbage bags to U.S. Army tanks.

16. High-Tech Ceramics

Ceramic materials are hard, chemically inert, and resistant to corrosion, wear, and high temperatures. Any substance except carbon-based compounds can be used when making ceramic materials. Most ceramics are electrical insulators and are transparent to most forms of electromagnetic radiation.

17. Fiber-Reinforced Composites

Composites are materials such as ceramics and plastics that have been reinforced with synthetic fibers and carbon filaments. Composites are beginning to replace some automobile and airplane parts because they are lightweight, resist corrosion, and are often stronger than steel.

18. Superconductors

Superconductors are materials that carry electricity without any loss of energy. To be in a superconducting state, materials currently need to be cooled far below room temperature. The simple rule of superconducting economics is: The closer to room temperature the more valuable the superconductor.

19. Thin-Film Deposition

A process that deposits layers of a specific material as thin as a single atom onto almost any surface. One process called chemical vapor deposition (CVD) uses a coating material that is heated until it vaporizes and then allowed to condense into the surface to be coated. A second process, molecular beam epitaxy, uses a semiconductor fabrication process to build up devices one molecular layer at a time. This process allows different materials and types of doping to be sandwiched precisely in layers.

20. Molecular Designing

A technique for creating custom-made materials. Scientists using a supercomputer can decide what properties they want a material to have and then, using advanced computer graphics and modeling programs, custom-design a new material molecule-by-molecule, atom-by-atom. By using lasers to lay down atoms in a precise pattern on surfaces, molecular designers can alter the properties of materials, such as making metals become glass and insulators become conductors.

Twenty Core Technologies and Their Relationship to the Twenty-four New Tools

1. Genetic Engineering

  a. Recombinant DNA (rDNA) technology is the mapping, restructuring, and remodeling of the gene code to eliminate or enhance a specific trait.

  b. Antisense RNA compounds have the power to block the expression of specific genes. Many defects and diseases in both plants and animals can be eliminated by shutting off the defective gene.

2. Advanced Biochemistry

Although there are no qualifying new tools in this category at this time, there are significant areas that are benefiting us all.

  a. *Monoclonal antibodies* have been produced that bind only to a specific molecule and are used to diagnose disease, pinpoint specific genes, and purify rare substances.

  b. *Interleukin-2* is a class of drug that can fight diseases like cancer.

  c. *Fetal-cell transplants* can be used to treat blood disorders such as sickle-cell anemia, Parkinson's disease, diabetes, and radiation exposure. Fetal-brain tissue transplants have also been able to reverse memory loss and brain damage in test animals, which may lead to relief for people suffering from Alzheimer's disease, mental retardation, and injuries to the brain. Fetal-cell tissue has a great advantage over other tissue because it grows faster than adult tissue and is less likely to be rejected when transplanted.

d. *Photoactive drugs* allow therapeutic drugs to be placed in a specific location in the body and then activated when exposed to light.

3. Digital Electronics

a. Digital imaging converts paper-based data, such as words, pictures, charts, and graphics, into a digital form that computers can read and manipulate.

b. Digital interactive television uses a computer/processor and memory chips to process, store, create, and transmit information and programming. The viewer will shift from being a passive receiver to an active participant.

c. Digital cellular telephones have no distortion or hiss, are able to send computer information without a modem, and have a high level of security.

d. Personal communication networks (PCN) allow PCN digital telephones to operate much like cellular phones, but they are static free, far less expensive, and smaller.

e. Antinoise technology allows a digital sound-sampling device to sample any noise and then cancel it by generating an out-of-phase sound, which is a mirror-image of the original sound at the same volume as the original noise.

4. Optical Data Storage

a. Advanced compact disks contain large amounts of data, sound, and/or video in a relatively small space.

b. *Bar-code readers. See* Electronic data interchange (EDI).

c. *Optical film* records digital data using lasers much like a compact disk, except that the information is stored on long rolls of film.

d. *3-D holographic crystals* are a futuristic way to store digital data using lasers to place the information within the three-dimensional surface of a synthetic crystal. In the twenty-first century, this will become a major method for storing large amounts of data.

5. Advanced Video Displays

    a. Advanced flat-panel displays provide us with thin, black-and-white or full-color, lightweight television screens in a wide variety of sizes. Advanced flat-panel displays will become the monitor of choice for computers, location video advertising, and entertainment systems.

    b. *High-definition television (HDTV)* describes a very high resolution screen whose dimensions resemble those of a picture in a movie theater. HDTV displays will find their first applications in medicine and advanced computer simulations.

6. Advanced Computers

    a. Electronic notepads are hand-held computers that can be customized by the user to fit a particular need. They have pen-based input and an optional wireless communication system.

    b. Multimedia computers allow users to interact with integrated combinations of data, sound, and video.

    c. Telecomputer is a combination of a Touch-Tone telephone, fax, and computer with pen-based or touch-screen input using a flat-panel display connected to telephone lines.

    d. Parallel processing computers are computers that can solve complex problems quickly by using a large number of processors to simultaneously attack the problem.

    e. Multisensory robotics are computer-controlled machines that can be programmed to do a wide variety of tasks, move around, and react to their environment.

7. Distributed Computing

    a. Electronic data interchange (EDI) is a combination of computer hardware and software that allows different companies and industries to break through outdated boundaries by automating standard business transactions electronically instead of via traditional voice and paper methods.

    b. Desktop videoconferencing allows the user to see and hear the other parties they are connected to on their computer screen.

    c. Computer-integrated manufacturing (CIM) allows a manufacturing system's operations to fully integrate all functions from engineering and planning to the devices controlling the production line.

d. *Local area network (LAN)* is a network of computers and communications devices that share equipment covering a small geographic area; it enables users to communicate with each other without the need for a central processor.

e. *Wide area network (WAN)* is a computer network that uses a central processor in the form of a time-sharing minicomputer or large mainframe that links various shared devices through the telephone lines to cover a large, even worldwide, area.

8. Artificial Intelligence

a. Advanced expert systems (ES) are knowledge-based software programs that capture the expertise of decision makers, convert it to a set of rules, and apply those rules to problem solving.

b. Advanced simulations are computer programs that can be used to simulate objects or events. There are three basic types of advanced simulations: 2-D, 3-D, and virtual reality.

c. Object-oriented programming (OOP) enables new software programs to be built from prefabricated, pretested, picture-based building blocks of software code in a fraction of the time it would take to build a new program from scratch.

d. Fuzzy logic. All computers operate on a "yes-no" principle. Fuzzy logic is a software program that adds a "maybe," enabling computers to deal with higher levels of abstraction and handle conflicting commands.

e. Neural networks. Neural network computing goes beyond simply executing a fixed set of commands by *learning* complex patterns through training.

f. *Voice recognition* software allows computers to respond to the sound of your voice. Primary uses will focus on enabling the human voice to be the input device for computers.

g. *Image processing. See* digital electronics.

9. Lasers

A wide range of applications for lasers is already in use, including laser scalpels which are increasingly replacing the traditional metal scalpel in surgery, resulting in faster healing times. Two key applications are:

a. Advanced compact disks (CDs), which contain large amounts of data, sound, and/or video in a relatively small space.

b. *Holography* is a technique that uses lasers to produce 3-D imagery; it will be common in advertising, such as in a display window of a jewelry store.

10. Fiber Optics

a. *Fiber-optic telecommunications systems* can carry four signals at once: telephone, television, radio, and computer data.

b. *Distributed computing* systems can use fiber-optic links to facilitate the communication of large amounts of digital data.

c. **Endoscopic technology** uses an endoscope, which is a long, hollow tube containing a fiber-optic strand connected to a small video camera that can be inserted through hard-to-reach areas, thus enabling the user to see and manipulate objects that would otherwise be very difficult to reach.

11. Microwaves

Although there are no qualifying new tools in this category at this time, there are significant areas that are benefiting us all. Microwaves have two major application categories: Transmitting and receiving wireless digital information and heating objects by causing molecular movement inside the object. Uses range from microwave clothes dryers and microwave scalpels to heating inoperable cancerous tumors to a temperature that kills the cancer but leaves the healthy cells alive.

12. Advanced Satellites

Although there are no qualifying new tools in this category at this time, there are significant areas that are benefiting us all. Landsat, for example, is being used for oil and mineral exploration, and Navistar can be used to determine exact locations anywhere on the planet including the position of all forms of transportation. Surveillance will continue to play a big role for satellites. Other new examples include:

a. *Low Earth Orbit (LEO) satellites* will allow anyone on the planet with a digital cellular telephone to communicate with another phone user anywhere.

b. *Direct broadcast satellites (DBS)* transmit their broadcast signals through medium-power satellites using higher frequencies instead of the low-power satellites used by other services. This combination of stronger signals and higher frequencies means that the receiving antennas, or dishes, will be significantly smaller and less expensive. Benefits include a super-VHS-quality picture, CD-quality sound, capability of broadcasting HDTV signals, interactive menu-driven interface, and a much lower price.

13. Photovoltaic Cells (PV)

Although there are no qualifying new tools in this category at this time, there are significant areas that are benefiting us all. Currently, PV cells can convert sunlight directly into electricity at an efficiency rate of more than 28 percent. They can be used for applications such as pocket calculators, refrigerators, portable communications, and remote and rural electrification.

14. Micromechanics

Although there are no qualifying new tools in this category at this time, there are significant applications such as monitoring pollution, aiding medical research, and giving robots a sense of touch. *X-ray* or *electron-beam lithography* is a state-of-the-art process that etches lines as small as 20 nanometers—only 100 atoms—across. Future applications include the manufacturing of second-generation micromachines. *Quantum structures* are microscopic electronic structures, such as quantum walls, quantum wells, quantum wires, and quantum transistors, that will revolutionize the electronics industry in the early twenty-first century.

15. New Polymers

Although there are no qualifying new tools in this category at this time, there are significant areas that are benefiting us all.
  a. *Mixed-media polymers* can be used for airplane propellers, running shoes, compact disks, and low-friction ball bearings.
  b. *Conductive polymers* are materials that combine the electrical properties of metals with the advantages of plastic. They can be in the form of fiber films or other shapes. In 1987, conductive polymers matched the conductivity of copper, and research indicates that they may someday outperform copper as a conductor. There are already over twenty conducting polymers that

are lighter and stronger than transistors and more resistant to heat in circumstances where devices made from conventional silicon fail.

c. *Stereolithography* is a process that coalesces liquid polymers into highly complex solid structures in a matter of hours. This technique reduces the large amount of effort and expense required to produce prototypes of new designs.

16. High-Tech Ceramics

Although there are no qualifying new tools in this category at this time, there are many current applications such as abrasives for cutting tools, heat shields, ball bearings, engine components, and artificial bone implants, as well as everyday items like ceramic knives, scissors, and ballpoint pens.

17. Fiber-Reinforced Composites

Although there are no qualifying new tools in this category at this time, there are many current applications such as automobile and airplane parts because composites are lightweight, resist corrosion, and are often stronger than steel. Beech Aircraft Corporation's new business jet, the Starship, has an all-composite body.

18. Superconductors

Although there are no qualifying new tools in this category at this time, there are many current applications, such as less expensive but more advanced magnetic imaging machines for hospitals, superconducting television antennas, faster computer circuits in mainframe computers using thin-film superconductors, and small and efficient electric motors.

19. Thin-Film Deposition

a. *Chemical vapor deposition (CVD)* is a process that allows a coating material to be heated until it vaporizes and then condenses into the surface to be coated.

1) **Diamond thin-film coating** is a CVD process of depositing a layer of diamond film only several molecules thick on a surface like the blade of a knife, a razor, or even a loudspeaker, thus creating a surface that has the properties of natural diamond.

b. *Molecular beam epitaxy* is a semiconductor fabrication process used to build up devices one molecular layer at a time. This process allows different materials and types of doping to be sandwiched precisely in layers. Applications include a semiconductor laser fabricated onto a glass surface, improved optoelectronic integrated circuits, radiation-resistant electronics, and lower-cost solar cells.

c. *Ion implantation* is a coating process that bombards a base material with ions of the coating material so that some of them remain embedded in the surface.

d. *Ion beam assisted deposition* is a coating process that uses ions of noble gases to "nail" individual atoms of the desired coating material deep into a ceramic surface.

20. Molecular Designing

Although there are no qualifying new tools in this category at this time, there are many current applications such as custom designing of new materials molecule-by-molecule, atom-by-atom. The first products to move out of the lab are tailor-made enzymes for industry.

Twenty-four New Cards in the Game

1. **Electronic Notepads**

   This form of pen-based hand-held computer can be easily customized by the user to fit a wide variety of needs. Entering, storing,
   and retrieving information will be their primary functions. You will
   choose the configuration when you buy it with options such as:

   a. Black-and-white or color advanced flat-panel display.

   b. Wireless keyboard and, in the near future, voice recognition, to
   enter data when not using the standard pen-based system.

   c. Wireless communication system to:

   1) Communicate data to other computers using radio or cellular
   networks.

   2) Communicate voice or paging information to any telephone
   or pager anywhere.

   d. Flash-memory cards will be the data-storage medium of choice
   because of their light weight and low battery consumption.

   Note: Apple Computer calls its electronic notepad a personal
   digital assistant (PDA) and Apple's competitors will come up
   with their own terms to try to differentiate themselves. Don't let
   the variety of terms confuse you. If their primary functions are

entering, storing, and retrieving information, they will all fit under the electronic notepad classification.

2. Multimedia Computers

Multimedia combines the audiovisual power of the television, the publishing power of the printing press, and the interactive power of the computer. By blending personal computers with either digital video interactive (DVI) or CD-ROM-based storage systems, you can create the foundation for an interactive system that integrates data, sound, and video. Four main application areas for multimedia include:

a. Business presentations
b. Training and education
c. Reference database
d. Electronic correspondence

3. Parallel Processing Computers

Parallel processing involves an advanced computer-processing technique that allows a computer or a large number of processors to simultaneously attack a problem. Ten identical computers working together on a single problem will solve the problem much faster than a single computer working alone. Parallel computers can incorporate tens of hundreds or thousands of individual processors or, using parallel networking techniques, PCs can be linked to work in parallel. Parallel computing's biggest advantage is found in its ability to reduce the time required to retrieve and analyze data. The technology will allow users to deal with increasingly complex commercial and scientific tasks.

4. Advanced Compact Disks

Optical disks use lasers to read information that is stored in digital form. The public is familiar with the compact disk (CD) as the distribution medium of choice for music. The key benefits lie in the ability of the user to store, interact with, and retrieve large amounts of data in a relatively small space. The number of advanced compact disk types will continue to expand. The most familiar are:

a. *Compact disk read-only-memory (CD-ROM)*. CD-ROM data disks have been available for more than ten years but are now reaching a price level needed for mass acceptance. They will become widely used to replace instruction/user manuals, parts and supply catalogs, a wide variety of textbooks, and archival reference materials. CD-ROM jukebox players will allow users to quickly access large amounts of data. *CD-ROM XA* (the "XA" stands for extended architecture), is a more advanced version of a CD-ROM allowing additional data to be added to a CD-ROM disk.

b. *Compact disk interactive (CD-I)* consists of a playing device that interfaces with a television set, allowing the user to interact with the information without the need for a personal computer. The CD-I reader has a computer invisibly built in. The primary uses will be entertainment, education, and training. All CD-ROMs will eventually evolve into the CD-I format, giving the user an interactive option.

c. *Compact disk-recordable (CD-R)*. CD-R devices that will record information will become popular in the mid- to late 1990s as access time rivals that of a hard disk. The *floptical disk* is a CD-R that is the size of a small floppy disk. The disk drive will also accept current floppy disks.

d. The *photo CD* holds digital images from film-based photography.

5. **Digital Imaging**

Using combinations of a personal computer, a scanner, a CD, a compression processor, and an electronic camera, you can electronically capture, store, share, view, and manipulate photographic-quality images. Macroimaging will be used to convert paper-based data, such as words and graphics, into a digital form that computers can read and manipulate. Corporations have already begun to digitize all types of information, including written, audio, video, and film. This will allow for better organization and utilization of large amounts of information.

a. *Automated scanning systems* will allow organizations to convert large amounts of paper-based data to digital form quickly and efficiently.

b. *Hand-held and desktop scanners* will be widely used by individuals.

c. *Electronic filmless cameras* will increasingly be used to capture, store, share, view, and manipulate images.

Note: There are three basic categories of digital imaging:

1) *Macroimaging.* This is the type of digital imaging described above that I've referred to as a new tool.

2) *Microimaging.* Using MRI, CT scanning, ultrasound imaging, laser imaging, or teleradiation, you can create photographic-quality pictures of visible and invisible living and nonliving matter.

3) *Nanoimaging.* Using a scanning tunnel microscope (STM), you can create photographic-quality images of the invisible world, like the atoms in a molecule.

6. Advanced Simulations

Advanced simulations are software programs that can be used to simulate objects or events. There are three basic types of advanced simulations:

a. *2-D and 3-D simulations* are software programs that use high-power computers, high-resolution graphics, and digital-quality sound to graphically simulate various situations. For example, computer users can visualize the inside of a home that has not been constructed yet or see the environmental effects of a dam that has yet to be built. Simulations can be made of the unseeable, such as wind and temperature patterns around the earth or the molecular construction of an amino acid. Your home television is capable of displaying a video recording of advanced 2-D and 3-D simulations that were created on a computer system. To view a 3-D simulation, the user must wear special glasses that use different colored lenses or polarizing filters.

b. *Virtual reality* is a collection of computer hardware and software, including data gloves and a special helmet containing mini computer screens, that allows the user to step into virtual worlds simulated in three dimensions. The key near-term applications will be entertainment and training.

7. Advanced Expert Systems (ES)

Advanced expert systems are knowledge-based artificial intelligence (AI) software programs that capture the expertise of decision makers, convert it to a set of rules, and apply those rules to problem solving. They can be used for management decision support, sales support, accounting, personal decision making, and training. Using preprogrammed criteria, ES can put together enormous amounts of knowledge about a given subject and organize it so that the computer can render a decision by process of elimination. Expert systems comprise factual and heuristic knowledge. They are best used for routine, repetitive decisions where the knowledge base is easy to specify. Expert systems can be limited by a constantly changing environment, a set of rules that cannot be defined for uncertainty, time (taking up to several years to complete), and by costs that can range in the $200,000 area. Productivity benefits include time savings when working with a large number of variables, such as a manufacturing plant with 8,000 new manufactured items per year and a large number of new parts; accuracy by reducing errors when dealing with a large number of variables; and speeding the process of helping others utilize the captured expertise without having to log the years of training and experience of the original human experts.

8. Neural Networks

Neural network computing is a combination of both computer hardware and software whose capabilities are often referred to as biological, or brain-inspired, computing because it mimics the workings of the human brain. It does this by using an artificial neuron with multiple inputs that enables the software to learn through training and go beyond simply executing a fixed set of commands. It is most closely associated with pattern recognition. By storing data in layers, the computer can compare and evaluate information based on its fit with known data. This allows for image analysis and recognition, for converting handwritten information to digital data, and for assessing risk factors based on ambiguous data.

9. Object-Oriented Programming (OOP)

Object-oriented programming enables new software programs to be built from prefabricated, pretested, picture-based building blocks of software code in a fraction of the time it would take to build a new program from scratch. This will allow both computer programmers and laypeople to quickly customize software to specific needs.

10. Fuzzy Logic

All computers operate on a "yes-no" principle. Fuzzy logic is a software program that adds a "maybe." This allows software to operate at a higher level of abstraction and handle conflicting commands. Applications include subway trains that don't jerk when starting and stopping, and automatic transmissions that shift more smoothly.

11. Electronic Data Interchange (EDI)

Electronic data interchange is an excellent example of multiple-enterprise computer integration allowing different companies and industries to break through outdated boundaries by automating standard business transactions electronically instead of via traditional voice and paper methods. The benefits are increased accuracy, increased speed, and cost reductions across the board. EDI is changing business relationships by creating a paperless, speed-based interchange of digital information. A key enabling technology is *bar-code technology,* which provides a fast way to enter data into a computer, usually at the point of sale.

12. Computer-Integrated Manufacturing (CIM)

Computer-integrated manufacturing allows a manufacturing system's operations to fully integrate all functions from engineering and planning systems to the devices controlling the physical operations on the production line.

13. Multisensory Mobile Robotics

A robot is a computer-controlled machine that can be programmed to do a wide variety of tasks. Today they are primarily used for repetitive, dirty, and dangerous tasks, or tasks that require a high level of precision. The addition of multiple sensors and mobility is yielding a new generation of advanced robots. *Flexible manufacturing systems (FMS)* use various types of multisensory robots that can be quickly reprogrammed to assemble, weld, paint, or perform some other repetitive task on a new "short run" of products.

14. Digital Interactive Television

Your existing television will become interactive with the addition of a computer/processor device and memory chips that allow it to act like a computer with a graphic user interface. Instead of just receiving and displaying images, digital interactive television will process, store, create, and transmit. The viewer will shift from being a passive receiver to an active participant. This will change the delivery system of products and services as *virtual service corporations* are simulated on your home system. The next generation of televisions will be called "digital televisions," and they will have the computer built in. Using a neural network chip set, your television will be able to learn your viewing preferences by keeping track of the shows you watch.

15. Telecomputer

A telecomputer is a combination of a Touch-Tone telephone, fax, and computer with pen-based or touch-screen input using a flat-panel display connected to telephone lines. It will be used for services such as information-on-demand, fax-back-on-demand, purchasing items, and picture-phone functions. It will also send signals to and receive signals from your interactive television. Visual voice-mailboxes will be a popular feature.

16. Desktop Videoconferencing

Desktop videoconferencing is similar to a telephone conference call that includes live video. Each person uses a personal computer, a small video camera mounted on top of the computer monitor, and a broadband data transmission system that allows the user to see the other parties on the computer screen as they all interact with the information.

17. Advanced Flat-Panel Displays

Advanced flat-panel displays will provide us with black-and-white or full-color, flat, lightweight television screens in a variety of sizes. Their applications will vary from home television to advertising. There are three basic types:

a. *Advanced liquid crystal display (LCD)*
    1) Active matrix LCD
    2) Passive matrix LCD
b. *Thin-film transistor display*
c. *Vacuum microelectronic display (VMD)*. The VMD with the

most potential is the field emission display, one large vacuum
tube with millions of microscopic cold cathodes.

18. Personal Communication Networks (PCN)

PCNs are digital, ground-based networks of small radio transmit-
ters and receivers that allow PCN digital telephones to operate
much like cellular phones but are static free, far less expensive,
and smaller. Not all PCNs will be used for wireless telephone communications. The network will also be used for *personal data communicators,* small hand-held devices for customized data entry
and transmission applications.

Note: The primary difference between personal communicators
and PCN telephones and electronic notepads that have a wireless
communication option is that their primary purpose will be focused
on communications instead of data entry and storage.

19. Digital Cellular Telephones

Current cellular phones use analog electronics, and in large cities
the number of cellular subscribers is close to the maximum limit
because of analog limitations. Digital cellular telephones will soon
begin replacing analog cellular phones in large cities, allowing for
many more subscribers. In addition, digital cellular telephones
have no distortion or hiss, are able to send computer information
without modem, and have a high level of security.

20. Diamond Thin-Films

Diamond is the hardest substance known and can deal with power
and heat levels far beyond those that would destroy the electronic
properties of most materials such as silicon. By mixing methane and
hydrogen under the proper conditions, you get a diamond film
about a millionth of an inch thick that is as hard as natural diamond.
By depositing an ultra-thin layer of diamond film, only several
molecules thick, on a surface such as the blade of a knife, a razor,
or even a loudspeaker, you create a surface that has the properties
of natural diamond. Applications include scratch-proof windows
and glasses, longer-lasting sandpaper, electronic and infrared sensors, ultraviolet laser beams for communications, higher-quality
speakers, faster computer chips, and better razors and cutting tools.

21. Antinoise Technology

Using a digital sound-sampling device, any noise can be sampled and then canceled by generating a mirror-image, out-of-phase signal at the same volume as the original noise.

22. Recombinant DNA Technology

Recombinant DNA (rDNA) technology is the mapping, restructuring, and remodeling of the gene code to eliminate or enhance a specific trait. Agricultural applications will include crops that are insect proof, drought resistant, and nitrogen fixing. Both plants and animals will be engineered to produce and process a variety of drugs, industrial lubricants, and enzymes. Human applications will range from predicting inherited genetic diseases to applying gene therapy for correcting genetic disorders.

23. Antisense Technology

Antisense RNA compounds have the power to block the expression of specific genes. Many illnesses, such as many forms of cancer and inherited and infectious diseases, are caused when genes are expressed inappropriately. Gene therapy using antisense techniques will effectively shut off the genes that trigger the human illnesses. In the plant world, for example, antisense techniques can be used to shut off the genes that trigger tomatoes to turn mushy, allowing them to travel better and last longer on store shelves.

24. Endoscopic Technology

This technology can best be explained by describing a major application, endoscopic surgery. A long, hollow tube (endoscope) is inserted through an incision less than an inch long. Optical fibers inside the endoscope carry a bright light from outside the patient down to the tip of the endoscope. A tiny television camera at the viewing end of the endoscope, inside the patient, transmits pictures to a television screen, which the surgeon watches while performing the surgery. Patients recover in days rather than weeks, and swelling and pain are greatly reduced. By the end of the decade, 75 percent of all surgery will be minimally invasive using endoscopes. Joints, the abdomen, the heart, any body section can be operated on thanks to the endoscope. It is important to note that this type of surgery is often referred to by a confusing list of other names such as laparoscopy, usually the name of the primary cutting tool that is involved, which tends to be used exclusively in a specific region of the body. All of these minimally invasive procedures have one key tool that makes the entire procedure possible, the endoscope. Other examples of endoscopic applications include:

a. Submarines that use an endoscope to view the surface without having to use the larger periscope.

b. Mechanics and electricians who use an endoscope to view hard-to-reach areas.

The Shifting Focus

Of Corporate Culture

From	*To*
Status quo	Rapid change
Industry performance	Individual action
Incremental innovation	Fundamental change
Expansion	Consolidation
Sameness	Redirection
Corporate groups	Partnerships
New technology as a cost	New technology as a necessity
Cost/growth/control	Quality/innovation/service
Bottom line of last quarter	Global market share

Of Management

From	*To*
Management	Leadership
Cheerleaders	Visionaries
Focus on process	Focus on strategy
Manage by control	Manage by commitment
Decision by command	Decision by consensus
Accepting the status quo	Taking risks

Reacting to change	Initiating change
Managing today's crisis today	Managing tomorrow's opportunities today
Solving today's problems today	Solving tomorrow's problems today
Individual work	Teamwork
Controlling others	Empowering others
Negative reinforcement of bad behaviors	Positive reinforcement of good behaviors
Fixing the blame	Fixing the problem
Taking credit	Giving acknowledgments
Periodic improvement	Continuous improvement
Organization man	Migrant professional
Centralized decision making	Decentralized decision making
Reward and promote by seniority	Reward and promote by performance

Of Human Resources

FROM	*TO*
Focus on task	Focus on process
Job titles	Job skills
Individual values	Shared values
Isolated specialists	Multiskilled generalists
Work with your hands	Work with your brains
Workers' gloves protect hands	Workers' gloves protect product
Upgrading technology	Upgrading people
Periodic training	Just-in-time training
Job security	Job adaptability
Guarantee your employment	Guarantee your employability
Organization man	Migrant professional
Retirement at age 65	Re-engagement several times in a life

Of Price vs. Speed

FROM	*TO*
Price	Speed
Pay for products	Pay for time
Value material wealth	Value free time

Of Information

From	*To*
Access to capital	Access to information
More information	Focused information
Static information	Dynamic information
Automation and support	Integration and coordination
Focus on new technology	Focus on new applications of technology

Of Computers

From	*To*
Information Age	Communication Age
Collecting information	Sharing information
Words and numbers	Data and voice and video
Data processing	Decision processing
Fit user to interface	Fit interface to user
Nice to have (features)	Need to have (features)
Client server to mainframe	Client server using UNIX
Proprietary systems	Open systems
Gigabits	Terabits
Character interface	Graphic-user interface
Profits from hardware	Profits from software
Programming by programmers	Programming by users
Repair national *infra*structure	Repair national *info*structure
Paper used for information storage	Paper used for information display

Of Manufacturing

From	*To*
Sell what they make	Make what sells
Premanufacture to anticipate sales	Manufacture when ordered
Predemand manufacturing	On-demand manufacturing
Mass production	Lean production
Large inventory	Just-in-time inventory
Long cycle times	Short cycle times

Mastery of the art of replication	Mastery of the art of innovation
Focus on what to make	Focus on how to make it
Quality manufacturing	Flexible manufacturing
Focus on quality	Focus on design
Design for assembly	Design for disassembly
Upgrade internal *infra*structure	Upgrade internal *info*structure
Build a better product	Build a better path to the customer
Employees as assemblers	Employees as problem solvers
Mass production (common products)	Customized mass production (common products with unique features)

Of Globalization

FROM	*TO*
Design for assembly	Design for disassembly
Foreign competition invades manufacturing	Foreign competition invades services
Thinking global	Being global
Focus on internal market	Focus on global market
Global competition	Global collaboration
Independence	Interdependence

Of Cultural Barriers to International Marketing and Their Shift

FROM (*Most culture bound*)	*TO* (*Least culture bound*)
Consumer products	Industrial products
Established product categories	New products and categories
Simple technology	Complex technology
Items used in home	Items used away from home

National Focus: United States vs. Japan and Germany

UNITED STATES FOCUS:	*JAPAN AND GERMANY FOCUS:*
Financially driven	Large industry groups
Short-term focus	Long-term focus
Antagonistic toward rivals	Collaboration with rivals

Industry Shift in Focus

Law

From   *To*

Litigation   Mediation and arbitration

Logistics

From   *To*

Goods   Services

Television News

From   *To*

Report on something that has   Report on something that is
   happened   happening

Environment

From   *To*

Town dumps   Regional landfills
Dumping   Waste reduction and recycling

Electronic Data Interchange (EDI)

From   *To*

Big-unit shipments   Small-unit shipments
Managing inventory   Managing information

Thirty New Rules

The new rules used in this book:

1. If It Works, It's Obsolete.
2. Past Success Is Your Worst Enemy.
3. Learn to Fail Fast.
4. Make Rapid Change Your Best Friend.
5. See the New Big Picture (because technology alters reality).
6. Solve Tomorrow's Predictable Problems Today.
7. Think Ten Years Out and Plan Back to the Present.
8. Build Change into the Plan or Product.
9. Focus on Your Customer's Future Needs (based on the new big picture).
10. Sell the Future Benefit of What You Do.
11. Build a Better Path to the Customer.
12. Give Your Customers the Ability (to do what they can't do, but would have wanted to do, if they only knew they could have done it).
13. Time Is the Currency of the '90s.
14. Leverage Time with Technology.
15. Enter the Communication Age.
16. Render Your Cash Cow Obsolete (before others do it for you).
17. Upgrade Technology and Upgrade People.

18. Creatively Apply Technology.
19. Network with All.
20. Re-become an Expert.
21. Find Out What the Other Guy Is Doing and Do Something Else.
22. Change the Way People Think.
23. Develop Collaborative Interactions.
24. Don't Fix the Blame, Fix the Problem.
25. Re-invent Successes of the Past Using the New Tools.
26. Re-invent Failures of the Past (as the successes they were meant to be).
27. High Touch Means High Cost.
28. Use Old Technology in New Ways.
29. Once a Card Is in the Deck, It Will Be Played.
30. Take Your Biggest Problem and Skip It.

Nine Revolutions the New Tools Will Create

1. Revolutionize the delivery system of products and services:
   a. Telecomputer
   b. Advanced Compact Disks
   c. Electronic Data Interchange (EDI)

2. Revolutionize the way we communicate:
   a. Personal Communication Networks (PCN)
   b. Desktop Videoconferencing
   c. Digital Cellular Telephones

3. Revolutionize the way we use and view television:
   a. Digital Interactive Television
   b. Advanced Flat-Panel Displays

4. Revolutionize the acceptance and use of computers:
   a. Electronic Notepads
   b. Advanced Expert Systems
   c. Object-Oriented Programming (OOP)

5. Revolutionize the way we individualize and personalize education:
   a. Multimedia Computers

6. Revolutionize the way we internalize, understand, and use massive amounts of data:
   a. Parallel Processing Computers
   b. Advanced Simulations
   c. Digital Imaging

7. Revolutionize medicine and agriculture:
   a. Recombinant DNA Technology (rDNA)
   b. Antisense Technology
   c. Endoscopic Technology

8. Provide the foundation for revolutionary new products and variations of old products:
   a. Fuzzy Logic
   b. Neural Networks
   c. Diamond Thin-Films
   d. Antinoise Technology

9. Revolutionize manufacturing:
   a. Computer Integrated Manufacturing (CIM)
   b. Multisensory Mobile Robotics

Suggested Reading List

Books

The collection of books listed below will help you to see the new big picture from other perspectives.

Aukstakalnis, Steve, and Blatner, David. *Silicon Mirage: The Art and Science of Virtual Reality*. Peachpit Press, 1992.

Brennan, Richard P. *Levitating Trains and Kamikaze Genes: Technological Literacy for the 1990s*. HarperPerennial, 1990.

Clarke, Arthur C. *Profiles of the Future*. Rev. ed. Warner, 1985.

Davidson, Frank P., and Meador, C. Lawrence. *Macro-Engineering: Global Infrastructure Solutions*. Ellis Horwood/Simon & Schuster, 1992.

Davis, Stan, and Davidson, Bill. *2020 Vision: Transform Your Business Today to Success in Tomorrow's Economy*. Fireside/Simon & Schuster, 1991.

Diebold, John. *The Innovators: The Discoveries, Inventions, and Breakthroughs of Our Time*. Plume, 1990.

Drexler, K. Eric. *Engines of Creation*. Anchor Press, 1986.

Drexler, K. Eric; Peterson, Chris; with Pergamit, Gayle. *Unbounding the Future: The Nanotechnology Revolution*. Morrow, 1991.

Drucker, Peter F. *Managing in Turbulent Times*. Harper & Row, 1980.

Dychtwald, Ken, and Flower, Joe. *Age Wave: The Challenges and Opportunities of an Aging America*. Bantam, 1989.

Fisher, Jeffrey A. *Rx 2000: Breakthroughs in Health, Medicine, and Longevity*. Simon & Schuster, 1992.

Howard, William G., Jr., and Guile, Bruce R. *Profiting from Innovation: The Report of the Three-Year Study from the National Academy of Engineering*. The Free Press/Macmillan, 1992.

Kuhn, Thomas S. *The Structure of Scientific Revolutions*. University of Chicago Press, 1970.

Kurzweil, Raymond. *The Age of Intelligent Machines*. The MIT Press, 1990.

Lee, Thomas F. *The Human Genome Project: Cracking the Genetic Code of Life*. Plenum, 1991.

Leebaert, Derek. *Technology 2001: The Future of Computing and Communications*. The MIT Press, 1991.

Morrisey, George. *Creating Your Future: Personal Strategic Planning for Professionals*. Berrett-Koehler, 1992.

National Research Committee on Research Opportunities in Biology. *Opportunities in Biology*. National Academy Press, 1989.

Osborne, David, and Gaebler, Ted A. *Reinventing Government: How the Entrepreneurial Spirit Is Transforming the Public Sector*. Plume, 1992.

Rheingold, Howard. *Virtual Reality*. Summit Books, 1991.

Rossman, Parker. *The Emerging Worldwide Electronic University: Information Age Global Higher Education*. Greenwood Press, 1992.

Strauss, William, and Howe, Neil. *Generations: The History of America's Future 1584 to 2069*. Quill/Morrow, 1991.

Theobald, Robert. *Turning the Century: Personal and Organizational Strategies for Your Changing World*. Knowledge Systems, Inc., 1992.

Toffler, Alvin. *The Third Wave*. Morrow, 1980.

Magazines

This short list of periodicals will provide an ongoing source of information to stimulate your creativity. All are published monthly unless otherwise indicated.

Discover
The Futurist (bimonthly)
High-Technology Business
Popular Science
Robotics Age (bimonthly)
Science

Science Digest
Science News (weekly)
Scientific American
Technology Review (bimonthly)

Specialized Publications

This list of highly focused publications represents only a small fraction of the technical information available to you. To keep you from drowning in a sea of technical information, I have selected a few that will help you quickly zero in on new opportunities.

Bioelectronics, published by Foster & Sullivan Technology Impact Report, 106 Fulton St., New York, NY 10273.

Corptech Technology Directory, published by The Planning Forum, P.O. Box 70, Oxford, OH 45056.

4,000 Abstracts "On Disk," published by Strategic Intelligence Systems. Tel. 212-605-0535.

Future Survey, published monthly by World Future Society, 4916 St. Elmo, Bethesda, MD 20814.

Oxbridge Directory of Newsletters, 150 Fifth Ave., Suite 301, New York, NY 10011; contains 16,500 U.S. and Canadian newsletters under 168 subject categories. Sample newsletters and reports that are available:
AI Trends
Applied Artificial Intelligence Reporter
Applied Genetics News
The Cambridge Report on Superconductivity
Coal and Synthfuels Technology
Fiber-Optics Sensors and Systems
Materials and Processing Report
Semiconductor Economics Report
Space Business News

Phillips Publishing, 7811 Montrose Rd., Potomac, MD 20854; publishes a variety of technology reports and newsletters such as:
Fiber-Optics News
Military Fiber-Optics News
Satellite News

Robotics, published by Elsevier Science Publishers, P.O. Box 201, 1000 AE, Amsterdam, The Netherlands.

Technotrends Newsletter, published monthly by Burrus Research Associates, Inc., P.O. Box 26413, Milwaukee, WI 53226. Tel. 800-827-6770.

U.S. Department of Commerce, National Technical Information Service, 5285 Port Royal Rd., Springfield, VA 22161; publishes a variety of reports and newsletters such as:

Foreign Technology Abstract Newsletter

Manufacturing Technology Abstract Newsletter

Tech Notes

Sources

5. Picture Power

Stewart, Thomas A. "Brainpower." *Fortune*, 3 June 1991, pp. 44–45. A report on IDS Financial Service's experience with advanced expert systems.

Op. cit., p. 50. A reference to Ford Motor Company's successful use of an advanced expert system.

Op. cit., p. 54. A discussion of corporate investment strategies.

6. From the Edge of Disaster to the Cutting Edge

Davidow, William H., and Malone, Michael S. *The Virtual Corporation*. New York: HarperCollins, 1992, pp. 121-22. Computer power and speed.

Shioya, Yoshio. "Genetically Engineered Foods Sprouting All Over the Globe." *The Nikkei Weekly* [Tokyo], 13 June 1992, p. 521. Figures on genetically engineered plants.

Burrus, Daniel, and Thomsen, Patti. *Advances in Agriculture*. Dubuque, Iowa: Kendall/Hunt Publishing Company, 1990. Recombinant DNA.

Weintraub, Pamela. "The Coming of the High-Tech Harvest." *Audubon,* July–August, 1992. Information on biotechnology, world population growth, and business opportunities.

7. This Way Out

Keller, Mary Ann. *Rude Awakening: The Rise, Fall and Struggle for Recovery of General Motors.* New York: William Morrow, 1989. Figures on GM modernization.

10. Toward Higher Education

CNN Headline News, March 1993. U.S. dropout rate.

Henkoff, Ronald. "Where Will the Jobs Come From?" *Fortune,* 19 October 1992. Examples of the tightening U.S. job market.

————. "Companies That Train Best." *Fortune,* 22 March 1993. Regarding statistics on corporate spending on employee training.

11. Long-term Problems and Short-term Thinking

"A Confederacy of Glitches." *Newsweek,* 21 September 1992. The government's computer illiteracy.

"Superservers May Make Someone Super-rich." *Business Week,* 24 August 1992. Background on superservers.

Gabor, Andrea. *The Man Who Discovered Quality.* New York: Penguin Books, 1990. Information on the National Commission on Productivity.

Branscomb, Lewis M. "Does America Need a Technology Policy?" *Harvard Business Review,* March–April, 1992. Supply side vs. the demand side of R&D.

12. Whole Life

Success, September, 1991. Results of a study of the Yale class of 1953 regarding planning for the future.

James Gleick, author of *Genius: Richard Feynman and Modern Physics,* interviewed on WAMU-FM, Washington, DC, 21 October 1992. About Feynman's power of observation.

Wilson, Douglas L. "Thomas Jefferson and the Charter Issue." *Atlantic Monthly,* November, 1992. Regarding Henry Adams and the theory of acceleration.

Robert Reich, interviewed on *The MacNeil-Lehrer NewsHour,* 21 October 1992. Comments on U.S. R&D levels.

Index

About the Author

Daniel Burrus, considered one of the world's leading science and technology forecasters, is the founder and CEO/president of Burrus Research Associates, Inc., a research and consulting firm that specializes in global innovations in science and technology, their creative application, and future impact.

His firm uses specially developed information-gathering techniques that enable it to help corporations, associations, and universities creatively apply cutting-edge information as they develop both short- and long-range plans.

His company mission is to help people of all ages learn to understand and profit from technological change and to become excited and involved in building a better tomorrow by discovering creative uses for the new tools of science.

His client list includes a wide range of industries, including many Fortune 500 companies, such as Motorola, 3M, AT&T, GE, Exxon, Du Pont, IBM, Hewlett-Packard, NCR, and Philip Morris.

Mr. Burrus produces a variety of publications, including *Technotrends Newsletter,* and is the author of several audio- and videocassette learning programs. He is the coauthor of five books, including *The New Tools of Technology, Advances in Agriculture, Medical Advances, Environmental Solutions,* and *Insights into Excellence*.

His interest in research became apparent in his third year of college, when he became one of the first undergraduates in the nation to direct a federal research grant.

He graduated with a B.S. degree and began his professional life as an innovative science instructor.

He went on to found and manage five businesses in a variety of fields, one of which, in the field of aviation, was a national leader with more than thirty locations.

Mr. Burrus is one of the most sought-after speakers on the international speaking circuit, delivering more than a thousand speeches in the past ten years to top executives, salespeople, and educators around the world. He is one of fewer than a hundred people worldwide to have been honored with the National Speakers Association's highest award for speaking skills and professionalism, the CPAE.

When not consulting, speaking, or writing, he plays several musical instruments and has a recording studio where he composes and records original works. He is an accomplished fine-arts nature and wildlife photographer and has won several local independent filmmaking awards. He built and flew several experimental aircraft and loves to snorkel, scuba dive, backpack, and travel to distant lands.

Technotrends™ Card Pack

Available at leading bookstores or from the publisher.

U.S. Games Systems, Inc.
179 Ludlow Street
Stamford, CT 06902
Telephone: (203) 353-8400
Toll-free: 1-800-544-2637
Fax: (203) 353-8431

Technotrends™ Video, Multimedia, and Audio Learning Systems

Available from:

Burrus Research Associates, Inc.
P.O. Box 26413
Milwaukee, WI 53226
Telephone: (414) 774-7790
Toll-free: 1-800-827-6770
Fax: (414) 774-8330

Special Offer

Receive a *free* trial subscription to the *Technotrends*™ *Newsletter.* Write or call Burrus Research Associates, Inc., and state that you have purchased the book *Technotrends.*

If you would like to share any comments, experiences, or ideas about *Technotrends,* please write to me at the following address:

Daniel Burrus

Burrus Research Associates, Inc.

P.O. Box 26413

Milwaukee, WI 53226